RAND NATIONAL DEFENSE RESEARCH INSTITUTE

T0303214

Advancing the Careers of Military Spouses

An Assessment of Education and Employment Goals and Barriers Facing Military Spouses Eligible for MyCAA

Esther M. Friedman, Laura L. Miller, Sarah E. Evans

Prepared for the Office of the Secretary of Defense

For more information on this publication, visit www.rand.org/t/RR784

Library of Congress Cataloging-in-Publication Data is available for this publication.
ISBN: 978-0-8330-8865-9

Published by the RAND Corporation, Santa Monica, Calif.

Preface

The My Career Advancement Account (MyCAA) is a Department of Defense (DoD) financial assistance program designed to help eligible military spouses pursue associate's degrees, occupational certificates, or licenses in portable career fields. The current version of the program dates from 2010.

As part of an effort to understand the value of MyCAA and to identify options for improving it, the Office of the Deputy Assistant Secretary of Defense for Military Community and Family Policy (MC&FP) asked the RAND Corporation to analyze data from DoD's 2012 Active Duty Spouse Survey (2012 ADSS) to learn about the education and employment goals and experiences of military spouses eligible for MyCAA. This research is part of a larger RAND project to address the design, monitoring, and evaluation of MC&FP's portfolio of Spouse Education and Career Opportunities (SECO) programs and initiatives. MC&FP sponsored the 2012 ADSS, which it developed with the Defense Manpower Data Center (DMDC).

This research should be of interest to policymakers responsible for programs or oversight of programs supporting military spouse quality of life, as well as scholars who study military spouse issues. It may also interest scholars who study education and employment benefits more generally.

This research was conducted within the Forces and Resources Policy Center of the RAND National Defense Research Institute, a federally funded research and development center sponsored by the Office of the Secretary of Defense, the Joint Staff, the Unified Combatant Commands, the Navy, the Marine Corps, the defense agencies, and the defense Intelligence Community.

For more information on the RAND Forces and Resources Policy Center, see http://www.rand.org/nsrd/ndri/centers/frp.html or contact the director (contact information is provided on the web page).

Contents

Figures and Tables

Figures

Tables

Summary

Since the move to an all-volunteer force, the U.S. military has increasingly recognized the importance of considering the well-being of families of military service members. The military services currently provide an array of services to support military families, including programs to assist spouses in pursuing their education and employment goals, child care assistance programs, and many others. Education- and employment-related programs are particularly important, given that military spouses face several unique challenges related to military life that can make it difficult for them to maintain and develop careers.

One program designed to help military spouses meet their educational and employment objectives is the U.S. Department of Defense's (DoD's) My Career Advancement Account (MyCAA) scholarship program. MyCAA is an educational financial assistance program that provides up to $4,000 in tuition and examination assistance for eligible spouses pursuing associate's degrees, occupational certificates, or licenses in portable career fields. The current version of the program began in October 2010.

As part of an ongoing effort to understand the value of MyCAA and to identify options for improving it, the Office of the Deputy Assistant Secretary of Defense for Military Community and Family Policy (MC&FP) asked RAND to analyze data from DoD's 2012 Active Duty Spouse Survey (2012 ADSS). The objective was to learn about the education and employment preferences and experiences of military spouses eligible for MyCAA and to identify barriers to MyCAA use and to achieving educational and employment goals more broadly. RAND's analyses concentrated on the population eligible for MyCAA: the 4,454 respondents who reported that they are not military personnel themselves and who are, according to military personnel records, spouses of active-duty service members in pay grades E-1 to E-5, W-1, W-2, O-1, or O-2.

Use of MyCAA

We first sought to understand who is and who is not using the program. Among those not using the program, we sought to understand the reasons for nonuse.

The survey showed that nearly one in five eligible spouses (18 percent) used MyCAA in the previous year. Of those who did not use MyCAA, 54 percent reported that they were unaware of the program.

Demographic factors differed considerably among those who had used MyCAA in the previous year and those who had not. Recent MyCAA users were younger than rank-eligible nonusers, more likely to be female, and less likely to be white. Recent MyCAA users were much more likely than nonusers to have some college (but no college degree), and they were much less likely than nonusers to have a bachelor's degree. Recent MyCAA users were also significantly more likely than nonusers to have earned a vocational or technical certificate. MyCAA users were also more likely than MyCAA nonusers to be married to a service member in the Army (50 percent compared to 44 percent) and were more likely to be married to personnel in junior enlisted pay grades of E-1 to E-4 (62 percent compared to 55 percent of nonusers).

Demographic factors also differed considerably between nonusers who were aware of MyCAA and those who were not. Nonusers who were unaware of MyCAA differed from nonusers who were aware of the scholarships in a number of statistically significant ways. For example, they were much more likely to have been married for less than three years. Spouses unaware of MyCAA were also more likely than other nonusers to live in civilian housing, to be men, and to be married to junior enlisted personnel, and they were less likely to be married to noncommissioned officers (NCOs).

Key reasons for not using the MyCAA program among those who were aware of it included concerns about eligibility and availability of time for education. Seven percent of these rank-eligible spouses who knew about MyCAA but did not use it in the previous year thought that their service member's rank was ineligible for MyCAA. Most of those spouses were married to E-5s.

The most frequently reported reason for not using the program in the previous year, among spouses who were aware of the program, was time constraints. This is consistent with civilian employer educational assistance programs, which also report time as the biggest barrier to program utilization.

Higher Education

MyCAA is designed to help eligible spouses establish and reach educational goals for associate's degrees, occupational licenses, and certificates in high-growth, high-demand career fields. To understand how well the program is helping spouses achieve those goals, we examined the percentage of spouses who are in school and, for those who are not, whether they want to be; and, for spouses who are not in school but want to be, the reasons why they are not pursuing additional training or schooling.

Recent MyCAA users were more likely than nonusers to be in school. As shown in Figure S.1, 28 percent of spouses were enrolled in school or training at the time of the survey and many more would like to be: 46 percent were not enrolled but would like to be. However, only 25 percent of recent MyCAA users would like to be in school but were not; that figure is statistically significantly lower than the percentage of spouses who did not use MyCAA and who wished they were in school (25 percent compared to 51 percent, respectively).

Nearly half of all rank-eligible spouses would have liked to be in school but were not. Our analyses of the education level of these spouses found that 55 percent reported having completed some college (but no degree) or an associate's degree. Given the education levels sponsored by MyCAA scholarships, it is not surprising that many more MyCAA users fall into these two education levels than do nonusers (73 percent compared to 53 percent). Nineteen percent of spouses who would like to be in school reported a high school diploma or equivalent as their highest education level, and 20 percent of them reported a bachelor's degree, compared to only 11 percent of MyCAA users overall who fell into the high school degree or equivalent category and only 13 percent of whom have a bachelor's degree.

In addition, 22 percent of spouses who wanted to be in school but were not already hold a vocational degree. Spouses with the desire for additional education or training outside MyCAA eligibility may need help connecting with other sources of financial aid to meet their needs.

Cost was the key reason given for not pursuing education. Spouses who wanted to be in school but were not most commonly reported the cost of education as a reason—this was cited by about 82 percent of all respondents. This finding demon-

Figure S.1
Percentage of Survey Respondents Enrolled in School or Training

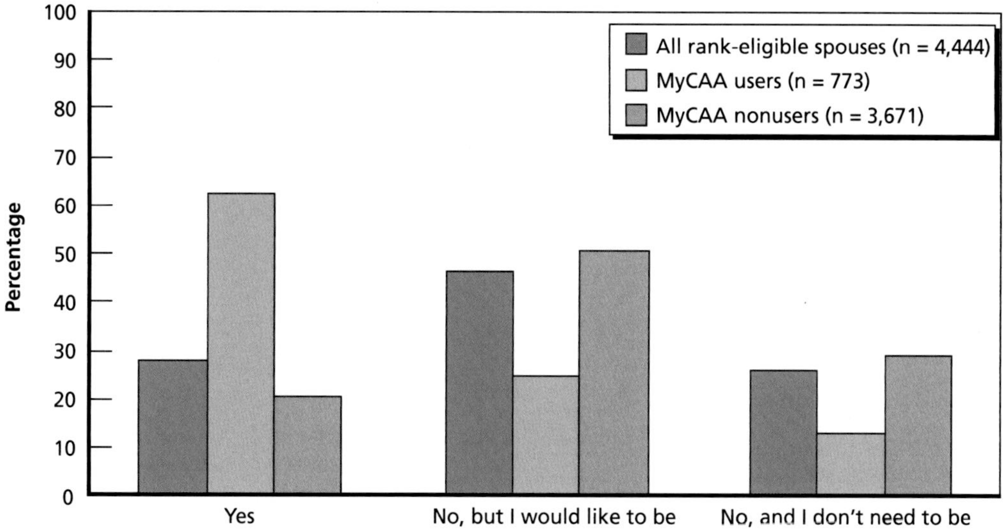

RAND *RR784-S.1*

strates that the MyCAA program is targeting a current need by providing a way for eligible spouses to address some of the costs of education.

Other barriers to education included family responsibilities and the cost of child care. Family responsibilities and the cost of child care were educational hurdles shared by both recent MyCAA users and spouses who had not used MyCAA in the previous year. Other, less-common barriers included transportation problems and inconvenient school hours.

Employment

MyCAA is designed to help military spouses obtain the education and training that will serve them well in the job market and reduce unemployment and underemployment among this population. Thus, it is important to examine the extent to which MyCAA users and nonusers are working, why they may not be working, and whether they are successfully employed in jobs that are a good match for their training and are satisfying.

About 40 percent of spouses in the survey reported that they worked for pay or profit in the prior week. Among the subset of respondents in RAND's analyses, 39 percent of all spouses, and 34 percent of MyCAA users, worked for pay or profit in the week before the survey.

The vast majority of spouses not looking for work (70 percent) said that they wanted to work. Thirty-seven percent of spouses who were not looking for work reported that they need to work. However, not all spouses wished to be working outside the home. About one-third of rank-eligible spouses indicated that they were not working and looking for work. Key reasons spouses do not look for work included a preference to stay home with children (73 percent) and the expense of child care (62 percent).

Recent MyCAA users and nonusers differed significantly in terms of employment. MyCAA users were significantly less likely to be working than spouses who did not use MyCAA in the past year, possibly because the former are in school. MyCAA users were also significantly less likely than nonusers to be looking for work and significantly more likely to want to be working.

Many working spouses were underemployed. Almost one-third (29 percent) of spouses were working part-time because they could not find full-time work. Other factors behind spouses' working part-time also included desires to spend time with family and child care considerations.

Recent MyCAA users were less likely to be employed in their field of education and required more time to find work following a move. Employed MyCAA users were less likely than employed nonusers to be employed working in their field of education but were much more likely than nonusers to be in school at the time of the survey. MyCAA users tended to take more time finding work after military moves than nonusers and took more time to acquire a new professional or occupational license after a military move. These could be the experiences that drew these spouses to the MyCAA program, as they are problems MyCAA is designed to help address.

Recommendations

RAND developed recommendations designed to address the issues raised by the analyses of the 2012 ADSS data.

Use

Help spouses manage competing responsibilities so that they can benefit from MyCAA scholarships. Career counselors at the Military Spouse Career Center might help spouses find ways to manage competing responsibilities (e.g., by linking them to child care options or connecting them to other staff at Military OneSource for assistance in locating a home repair referral source).

Promote MyCAA on an ongoing basis. MyCAA's primary targets are the spouses of service members who are entry-level or very early in their careers. The survey showed that more than half of rank-eligible nonusers were unaware of MyCAA. These unaware nonusers were much more likely to have been married less than three years than nonusers who were aware of the scholarships. Because there is a continuous flow of new spouses into the military community at the MyCAA eligibility level, promotion of MyCAA must be ongoing.

Coordinate with the four services to ensure that promotional activities for MyCAA target all eligible pay grades. DoD needs to coordinate with the services to ensure that MyCAA is not mistakenly advertised as being solely for junior enlisted spouses or for spouses of "junior personnel" and to ensure that promotional activities target spouses of all eligible pay grades. To promote understanding of eligibility among spouses, DoD could also describe eligibility in terms of service-specific rank (e.g., corporal) and not just pay grade.

Education

Help spouses interested in postsecondary education identify financial aid opportunities. Spouses may be eligible for federal financial aid from the Department of Education. Service members with more than six years of service are eligible to transfer their Post-9/11 GI Bill educational benefits to their spouses. States, educational institutions, and private foundations are other sources of financial assistance for postsecondary education. Career counselors at DoD's Military Spouse Career Center are accessible through Military OneSource and are equipped to help spouses identify financial aid opportunities beyond MyCAA for spouses of any pay grade who are interested in pursuing any type of postsecondary education.

Have Military Spouse Career Center career counselors inquire about child care needs. This practice could contribute to DoD's efforts to ensure that spouses are aware of DoD-subsidized child care options and how to apply. The process for obtaining a MyCAA scholarship requires that new applicants speak with a career counselor from the Military Spouse Career Center, so DoD has the opportunity to ensure that every MyCAA recipient has been offered information about assistance with the cost of child care.

Investigate whether military child care options can be aligned to match the needs of spouses in school. DoD child care availability and access is designed primarily to meet the needs of employed spouses. DoD should consider investigating whether it would be feasible to align military child care options to also match the needs of spouses in school in terms of available hours and schedules that may shift by semester.

Investigate whether MyCAA partner institutions offering on-site child care would be willing to offer child care discounts, scholarships, or space priority to MyCAA recipients during their enrollment terms. Many schools do not offer on-site child care services for students—particularly private and for-profit institutions—and those that do may have waiting lists. Still, some spouses who attend schools that do offer child care might be able to benefit if discounts or scholarships could be made available or military spouses could receive some level of priority on any waiting lists. Schools might be motivated to offer MyCAA child care discounts if MyCAA gave the schools that offer this benefit additional visibility on the MyCAA website. Perhaps some schools would be willing to offer such a benefit to any enrolled military spouse. Some colleges already offer scholarships to help student parents with the cost of child care and may be willing to add "military spouse" to their eligibility criteria.

Through Military Spouse Career Center career counseling, explore whether online classes can help spouses with challenges accessing education. Spouses reported that factors that prevent them from attending school or training include frequent moves, service member deployment, lack of transportation, inconvenient school hours or locations, and school conflicting with work schedules. Career counselors can help spouses identify programs available completely or partially online. Of course, many occupations require hands-on learning (e.g., animal trainer, electrician, chef), but some of the required coursework may be possible to obtain online, and some occupations may be particularly well-suited to online education (e.g., those in information technology).

Employment

Reach out to recent MyCAA graduates to assist with job searches. The Military Spouse Career Center could reach out to recent MyCAA graduates to provide job search support: Its career counselors are equipped to help any military spouse with job searches, resumes, interview preparation, and connections with Military Spouse Employment Partners (MSEP).

Connect spouses with opportunities for internships in their desired career fields while in school. Even before MyCAA users finish their occupational schooling or training, there are ways to improve their employment prospects. Internships might not be the right choice for all spouses or occupations or for every school term. Where appropriate, spouses should be encouraged to consider internships while in school, potentially with MSEP companies to assist in moving them into jobs more closely aligned with their current training. This offers the potential to help in a number of ways. First, spouses gain some work experience in their field (21 percent of spouses who were not employed reported that lack of necessary work experience was a reason they were not looking for work). Second, spouses who are employed fall into the first priority category for available space for full-time, DoD-subsidized child care, so paid internships would help address a barrier to education (child care costs). Third, internships can lead to employment with the same company, but even if they do not (the company does not have openings or the spouse moves), internship supervisors can potentially serve as job references. Regarding internships with MSEP companies, these partners have pledged to offer transferable, portable jobs so that spouses do not have to lose seniority or experience an employment gap following a Permanent Change of Station move. These partners include small or regional businesses, global businesses, defense contractors, universities, federal agencies, and nongovernmental organizations. Finally, there should also be continued integration of the MyCAA, Military Spouse Career Center, and MSEP programs, which could help expand internship opportunities for MyCAA users while in school.

Have Military Spouse Career Center career counselors inquire about callers' child care needs. We echo our recommendation above: Leverage the moment when spouses consult with career counselors about suitable occupations, job searches, and other employment issues to offer information about DoD-subsidized child care.

Acknowledgments

Cathy Flynn and Meg O'Grady in the Office of the Deputy Assistant Secretary of Defense for Military Community and Family Policy (MC&FP) approved and facilitated access to the survey data analyzed in this report and provided a great deal of contextual information and feedback on briefings of these survey results, conclusions, and recommendations. (Meg has since left this office.) We would also like to acknowledge Pat Shanaghan at MC&FP and Spencer Mann at BAM Technologies for support with MyCAA program data. Additionally, we thank Mike DiNicolantonio and other key staff at the Defense Manpower Data Center who also facilitated timely data access, provided material necessary for understanding the survey methods and dataset, and offered feedback on a prior study summary. Lee McMahon, Pat Shanaghan, and Yuko Whitestone reviewed a previous version of this report, provided useful suggestions, and confirmed MyCAA program details.

The authors would like to thank Gabriella Gonzalez for feedback throughout the research process and Katharina Best for assistance in cross-checking the data analyses. We would also like to acknowledge Paul Heaton, Larry Hanser, Jennifer Lewis, Sarah Meadows, and Jennie Wenger, whose constructive reviews of previous draft products from this work helped to strengthen this document. Paul Heaton at RAND and Mady Segal, Professor Emerita at the University of Maryland at College Park, served as our peer reviewers: We thank them for their thorough reviews and constructive feedback. Our writing also benefited from the editorial support of Kristin Leuschner.

Abbreviations

ADSS	Active Duty Spouse Survey
DMDC	Defense Manpower Data Center
DoD	U.S. Department of Defense
EFMP	Exceptional Family Member Program
GAO	U.S. Government Accountability Office
MC&FP	Military Community and Family Policy
MSEP	Military Spouse Employment Partners
MyCAA	My Career Advancement Account
NCO	noncommissioned officer
PCS	permanent change of station
SECO	Spouse Education and Career Opportunities

CHAPTER ONE

Introduction

Programs to support military spouses' higher education and employment are relatively new, as there was not much of a need for them in the draft era when most personnel were single and required to serve, or when women's workforce participation in the United States was very limited (Rostker, 2006, 2007; Miller et al., 2011b). Historically, U.S. military spouses did not receive specific accommodations or benefits of any kind. Military wives were expected not to be employed outside the home but instead to be homemakers, raise children, support their spouses, participate in military functions, and do volunteer work for the military (Segal, 1986). Until the late 1980s, officers' work performance was assessed in part on their wives' participation in volunteer activities both inside and outside the military. Spouses' activities were also documented in performance reports and used in determining whether a service member would be promoted (Harrell, 2000; Hosek et al., 2002). These expectations conflicted with any military spouses' interest in or need to develop their own careers and participate in nonmilitary hobbies and activities.

The advent of the all-volunteer force and changing social dynamics in the United States have led to the growing inclusion of women in the workforce and in the military (Rostker, 2006, 2007; Miller et al., 2011b). As spouse employment has risen, attention to their employment needs and the effect of military life on their employment and earnings have increased among policymakers (White House, 2011).

Today, the military services provide a wide array of programs, services, and facilities to support military families. Child care and employment assistance programs have often grown concurrently (Zellman et al., 2009). Other programs for military spouses include deployment assistance services, recreation programs, and Military OneSource—a defense-funded employee assistance program tailored to military life and available at no cost to military service members and their families. The White House declared military family support a national security policy priority in 2010 and highlighted spouse education and employment assistance programs as of particular interest (White House, 2011, p. 1).

Research has shown that military spouses tend to have more years of education than their civilian counterparts, yet they are generally less likely to be employed, and those who are employed tend to earn less, due in part to aspects of military life, such as frequent moves that disrupt employment, and depressed labor markets around military bases (Booth, 2003; Booth et al., 2000; Booth, Segal, and Bell, 2007; Cooney, 2003; Cooney, De Angelis, and Segal, 2011; Cooke and Speirs, 2005; Harrell et al., 2004; Heaton and Krull, 2012; Hisnanick and Little, 2014; Kniskern and Segal, 2010; Lim, Golinelli, and Cho, 2007; Little and Hisnanick, 2007).

Education- and employment-related programs are particularly important, given that military spouses face several unique challenges that make it difficult to maintain and develop careers. For example, discontinuous work histories limit the amount of work experience an individual can accumulate and may interfere with skill development and on-the-job training, which may dissuade employers from hiring (Cooney, De Angelis, and Segal, 2011). Moreover, those living in high-military labor markets may be disadvantaged in the job market, regardless of work history. For instance, Booth (2003) found that military wives living in labor markets with substantial percentages of active-duty personnel had lower returns to human capital investments, including work experience. Helping military spouses meet their education and employment goals has potentially important consequences for service member recruitment and military satisfaction and retention, as well as for the well-being of military families as a whole (Barnett, Gareis, and Brennan, 2008; Strazdins et al., 2010).

Recently, the U.S. Department of Defense (DoD) created a portfolio of Spouse Education and Career Opportunities (SECO). These programs include SECO career consultation services available through the Military Spouse Career Center, the My Career Advancement Account (MyCAA) program (described below), the Military Spouse Employment Partnership Program (MSEP), and DoD State Liaison Office initiatives to expand unemployment compensation eligibility to trailing military spouses (those following their service member after a permanent change of station [PCS] move) and to improve portability of occupational licenses and credentials.

The MyCAA Program

DoD's MyCAA program is an educational scholarship program that provides a lifetime maximum of $4,000 in tuition and examination assistance for eligible spouses pursuing associate's degrees, occupational certificates, or licenses in portable career fields. MyCAA is relatively new: The current version of the program dates back to October 2010. Although an earlier evaluation of MyCAA's program implementation documented successes and challenges associated with a pilot version of MyCAA, that evaluation occurred too early in the program's history to allow assessment of individual-level, short- and long-term outcomes. At the time of the study, most spouses had not completed their training or employment (Needels and Zaveri, 2009; Zaveri, Pisciotta, and Rosenberg, 2009).

The current version of MyCAA is the third major iteration of the program. An earlier demonstration version of MyCAA, which was implemented at 18 installations across eight states[1] beginning in 2007, provided spouses with grants of up to $3,000 a year for two years to use toward tuition, fees, or other relevant expenses at accredited two-year colleges and technical training centers or to use toward fees required for career credentials and licenses. As of May 2009, about 17 months after the selected states were to begin providing these scholarships, 5,366 MyCAA grants were awarded (Needels and Zaveri, 2009, p. 17).

An implementation study documented successes and challenges of the pilot program, and found in group discussions that spouses participated in MyCAA to

- further career and job prospects
- improve themselves and their self-esteem
- further their education
- contribute to family financial well-being
- engage in a positive distraction during their service member's deployment (Zaveri, Pisciotta, and Rosenberg, 2009, p. 57).

As noted above, this evaluation, while providing useful guidance for a full rollout of the program, was not able to assess program outcomes.

Following that demonstration program, a full-scale MyCAA tuition assistance program was launched across DoD in 2009 to promote employment and portable career opportunities by offering scholarships up to $6,000 per year to any military spouse. In less than a year, DoD was overwhelmed by the demand: 136,000 spouses enrolled in the program, for an estimated $250 million in benefits (Harkin, 2010, p. 6). The budget would not sustain a program on this scale, and MyCAA was closed to any new applicants. To save the program and contain costs, DoD scaled the program back by limiting eligibility, approved uses, and the amount of the scholarships (Harkin, 2010, p. 6).

Eligibility

The strategy for the redesigned program is to focus on the spouse population most likely to have the greatest need for assistance and on courses of study likely to provide a more direct and immediate benefit to the employment opportunities of military spouses. In October 2010, DoD launched a redesigned MyCAA program that restricted eligibility to spouses of military personnel in the earliest career ranks and pay grades. Thus, the current program targets spouses who are most likely to be early

[1] The states were California, Colorado, Florida, Georgia, Hawaii, Maine, North Carolina, and Washington.

in their education and career trajectories.[2] Spouses eligible for MyCAA must not be in the military themselves and must be married to service members who are serving on active-duty Title 10 orders and in the pay grades of E-1 to E-5, W-1 to W-2, and O-1 to O-2. These military personnel tend to be in their first term of service, so there is also the possibility that MyCAA could serve as a retention tool. Educational benefits from the Post-9/11 GI Bill cannot be transferred to spouses until military personnel have at least six years of service in the armed forces, so most of these spouses would not yet be able to acquire funds from that source.

Approved Uses

MyCAA scholarships focus on promoting "portable careers," a term used by DoD to refer to high-demand, high-growth careers identified by the Department of Labor as likely to be sustainable over time and that have job openings near military duty locations. MyCAA emphasizes these careers based on the following rationale:

> (Attainment of) a portable degree or credential funded with a CAA will help military spouses enter and advance in the workforce even as they relocate when their service member spouses transfer to other bases. A longer-term goal . . . is to encourage the retention of the service member in the military by improving his or her spouse's job prospects and increasing the satisfaction of the entire family with life in the military. (Zaveri, Pisciotta, and Rosenberg, 2009, p. 1)

MyCAA can be used to pay for education, training, and testing in hundreds of occupations falling into a diverse range of career fields. The original implementation of MyCAA specified only five career fields that met the program focus, including construction, education, financial services, health care, and information technology. In response to feedback from military spouses, MyCAA expanded the career fields offered to include human resources, hospitality, homeland security, and business administration (Needels and Zaveri, 2009, p. 24). This expansion has continued: A complete list of career fields and a sample of MyCAA-approved occupations are presented in Table 1.1. The consistent program goal across these changes has been sponsorship of the pursuit of careers that are portable to accommodate the military lifestyle and its frequent moves.

MyCAA funds can be used for tuition assistance for career-focused associate's degrees, occupational licenses, and certificates. MyCAA supports pursuit of a high school equivalency certificate or diploma only as a part of a plan to obtain one of these degrees, licenses, or certificates, not as an end in and of itself. The program does not support general studies or the pursuit of bachelor's or graduate degrees. Tuition for

---

[2] Most warrant officers are well advanced in their military careers, but for some Army aviators, this is the entry-level pay grade.

Table 1.1
Sample Portable Occupations, by Career Field

Career Field	Sample Occupations
Aerospace	Aircraft mechanic, aircraft service technician
Animal services	Animal groomer, animal trainer, veterinarian
Automotive services and transportation	Automotive mechanic, police/fire/ambulance dispatcher
Business, finance, and administration	Accountant, real estate agent, tax preparer
Construction	Carpenter, interior designer, painter, welder
Education	Child care worker, K–12 teacher, librarian
Energy	Electrical power-line installer and repairer
Health and human services	Dentist, dietician, medical billing, psychologist
Homeland security	Firefighter, police officer, security guard
Hospitality	Bartender, chef, lifeguard, event planner
Information Technology	Database administrator, software engineer
Legal	Court reporter, lawyer, paralegal
Skilled trades	Barber, journalist, nail technician, translator/interpreter

SOURCE: SECO, 2013.

graduate-level courses that are a part of an approved educational plan or for necessary continuing education in an approved career field may also be authorized.

In addition to tuition, MyCAA scholarships can cover occupational license or credential examinations and associated courses at any level, including at the bachelor's and graduate levels. For example, a teacher holding a bachelor's degree who needs professional recertification in a new state following a PCS move can use a MyCAA scholarship to pay for any courses or exams required for that certification. Many occupations and professions are regulated by state licenses and certificates: A few other examples not in Table 1.1 include truck driver, cosmetologist, electrician, licensed practical nurse, physical therapist, home inspector, computer network administrator, pharmacy technician, and human resource specialist.

Scholarship Amounts

Since October 2010, MyCAA scholarships have provided a maximum education benefit of $4,000, with an annual fiscal year cap of $2,000. Annual cap waivers are available for licensure and certificate programs if there is an upfront tuition cost that exceeds $2,000 (up to the maximum education benefit of $4,000). The funds are sent one school term at a time directly to the training or license/certification testing institution, not to spouses. MyCAA must approve the accredited institution or testing facility before granting the scholarship. If the cost of obtaining the degree, license, or certificate exceeds the $4,000 limit, a spouse must demonstrate the ability to cover the remaining expenses before an application will be approved. In addition to providing funding for schooling, the program connects spouses with career counselors at the Military Spouse Career Center who can help them develop an educational plan and access additional scholarship information, should additional funds be needed. Spouses must be able to finish their program of study within three years from the start date of the first MyCAA-sponsored course: Funds are no longer available after that expiration date. Spouses will also lose access to funds if, during this period, they become ineligible (e.g., through divorce, by joining the military themselves, or if their service member is promoted out of the eligible pay grade or leaves the military).

Use of MyCAA Scholarships

All eligible spouses who apply for a MyCAA scholarship for an approved course of study at an approved institution are provided with a scholarship (i.e., there is no set number of scholarships for which spouses must compete). Analyzing MyCAA program data provided by DoD, RAND found that more than 90,000 spouses have used MyCAA scholarships between the program redesign at the end of October 2010 and the end of November 2013.

Table 1.2 shows the estimated number of associate's degrees, certificates, and licenses obtained by military spouses using MyCAA scholarships in the three years since the program was reconfigured. We say "estimated" because in some cases the schools did not directly confirm completion of the educational plan; however, the school indicated that the final course or the occupational license or certification exam was successfully completed with a passing grade. These numbers likely underestimate completion, as spouses who lost MyCAA eligibility while still in school have no incentive to return to the MyCAA program system and record that they completed the education plan they started under the program.

Table 1.2
MyCAA Education Plans Estimated to Be Completed Between October 25, 2010, and November 26, 2013

Education Plan	Estimated Number Completed
Associate's degree	4,333
Occupational certificate/credential	19,286
Occupational license	1,076
Total	24,695

SOURCE: MyCAA program data provided by the Office of the Deputy Assistant Secretary of Defense for Military Community and Family Policy.

Focus of This Study

As part of an ongoing effort to understand the value of MyCAA and to identify options for improving it, the Office of the Deputy Assistant Secretary of Defense for Military Community and Family Policy (MC&FP) asked RAND to analyze data from DoD's 2012 Active Duty Spouse Survey (2012 ADSS) to learn about the education and employment goals and experiences of military spouses eligible for MyCAA and to identify barriers to involvement in MyCAA and to achieving educational and employment goals more broadly.

The 2012 ADSS complements MyCAA user data (e.g., occupational goals, funding, course enrollment, and completion) collected by the program for DoD internal monitoring of MyCAA program use. The analyses conducted in this study can help shed light on the value of MyCAA for spouses and identify ways to improve this program. In addition, the 2012 ADSS provides information on eligible nonusers of the program, who serve as a useful comparison group for MyCAA users. Given that the 2012 ADSS asks only about MyCAA use in the previous calendar year, this study may be thought of as an evaluation of recent MyCAA use and nonuse among rank-eligible spouses. The focus here is on short-term educational and employment outcomes—including attending school, whether spouses are working, and job characteristics and income.

The study examined the following questions:

- Who uses MyCAA?
  - Why do some spouses not use MyCAA?
  - How well does the program reach the population it is intended to support?
- What are the reported educational experiences, preferences, and barriers faced by MyCAA users and nonusers?
  - What percentage of spouses are in school or want to be but are not?
  - Why are some spouses not pursuing additional training and schooling, if they want to?
- What are the reported employment experiences, preferences, and barriers faced by MyCAA users and nonusers?
  - What percentage of spouses are looking for work?
  - Why are some spouses not looking for work?
  - Why do some spouses work part-time instead of full-time?
  - What are the job characteristics of employed spouses?
- What conclusions and recommendations can we draw from spouses' survey responses?

Organization of the Report

The remainder of this report describes the results of the RAND research team's analysis of the administration of the 2012 ADSS and resulting recommendations for the MyCAA program and for DoD's portfolio of SECO programs more broadly. Chapter Two describes RAND's overall study sample and methods. Chapter Three focuses on recent MyCAA use and barriers to use faced by spouses. Chapter Four investigates educational desires and outcomes of MyCAA users and nonusers. Chapter Five examines employment outcomes of MyCAA users and nonusers, including whether they are employed, reasons for not working, job characteristics, and underemployment. Chapter Six concludes the report with recommendations for ways in which DoD could potentially better assist military spouses with their education and employment trajectories. Appendix A describes the variables used in the analyses in Chapter Three, Appendix B compares MyCAA nonusers who are aware of the MyCAA program to those who are unaware, and Appendix C shows the results of multivariate models comparing user and nonuser characteristics, as outlined in Chapter Three.

CHAPTER TWO

Analytic Methods

This chapter describes the Defense Manpower Data Center's (DMDC's) administration and weighting procedures for the 2012 ADSS, and RAND's selection of the sample, survey items, and analytic approaches for this study.

DMDC's 2012 ADSS Administration and Survey Weights

DMDC and the MC&FP developed the 2012 ADSS (hereafter referred to as the survey), which was administered by DMDC. In fiscal year 2012, there were 709,776 active-duty spouses (U.S. Department of Defense, 2013, p. vii). The survey was open from November 19, 2012, through March 11, 2013, and about 13,000 spouses from all four services participated.

DMDC linked individual survey responses to demographic data from military personnel records, weighted to adjust for sample design and nonresponse. The weighted response rate was 23 percent. There is no set minimal response rate for a survey to be valid and no strong empirical relationship between response rates and nonresponse bias (Johnson and Wislar, 2012; Groves, 2006). Nonresponse bias occurs only when those who did not participate in the survey would have answered the questions differently than those who did. One way that survey researchers correct for potential nonresponse bias is to use information on characteristics that are expected to correlate with survey responses to weight responses so that the population of survey respondents in the analytic sample reflects the demographic composition of the full population of interest, in this case military spouses.

Fortunately, DoD's military personnel administrative files contain some key demographic information that can be used to prepare weights. Weights were provided to RAND by DMDC, which described the three-stage process for weighting the data:

> Analytical weights for the 2012 ADSS were created to account for unequal probabilities of selection and varying response rates among population subgroups. Sampling weights were computed as the inverse of selection probabilities. After determining case dispositions, the base weights were adjusted for eligibility, which

> was adjusted for completion to primarily account for nonresponse. The adjusted weights were poststratified to match population totals and to reduce bias unaccounted for by the previous weighting steps. (DMDC, 2013a, p. 10)

DMDC's methodological report also describes the sample design and selection, survey administration, steps for development of the final weights, and survey response rates (DMDC, 2013a). DMDC also provided unweighted demographics of survey respondents compared to the estimated population in its volume of tabulations for all participants (DMDC, 2013b, pp. 8–9).

Sixty-five percent of respondents selected for inclusion into this study took the survey online, and 35 percent responded to the paper version that DMDC mailed on December 10, 2012, to spouses who had not responded online by this date. Following up with spouses who do not participate online by mailing a paper survey is another legitimate strategy for addressing the potential for nonresponse bias, particularly if some spouses are less likely to have Internet access than others, and lack of access is linked to issues being assessed by the survey. For example, in the case of the 2012 ADSS, lack of Internet access could reflect financial difficulties (i.e., inability to afford Internet service), or it might hamper spouses' abilities to search for employment, take college classes, and use DoD SECO resources, such as MyCAA.

We compared paper and web respondents in our analytic sample of spouses, to be described in more detail below, and found several statistically significant differences between the two groups ($p < 0.05$) in terms of their demographic, family, and military service characteristics. For instance, paper respondents were statistically significantly more likely than web respondents to be female, younger, and white. Paper respondents were also more likely to be married to service members in the Army and Marine Corps and to junior enlisted personnel (E-1 to E-4).

By excluding the paper respondents and rerunning several key analyses on education and employment presented in this report, we found that the survey results would have been mostly comparable if DMDC had not made this follow-up effort. We did find a few noteworthy differences, however ($p < 0.05$). Although both web and paper respondents were equally likely to be employed, paper respondents were more likely to be working part-time rather than full-time, were less likely to be in school, and were less likely to be aware of MyCAA. Although the difference in percentage points is not great (3 to 5 percent), those differences are statistically significant. When we limited analyses to web respondents only, we find a greater percentage of MyCAA users reporting that conflicts with their work schedule prevented them from attending school or training (a 9-percentage-point difference). We did not find large differences when comparing both sets of respondents on other reasons for not attending school or training and reasons for not looking for work. In sum, if DMDC had relied solely on web participants for the 2012 ADSS, the results would have slightly underestimated the level of employment, college enrollment, and awareness of MyCAA among this population of rank-eligible spouses and to a greater degree underestimated how many spouses were able to deconflict work and school attendance.[1]

Sample Selected for RAND's Analyses

RAND's analyses concentrated on the population who could have been eligible to use MyCAA at the time of the survey. After excluding spouses who were nonrespondents or did not provide responses to whether they used MyCAA in the past year, we were left with a total of 13,017 spouse respondents. We excluded 2,358 spouses who reported on the survey that they were currently serving in the military and an additional 6,205 who, according to military personnel records, were not rank-eligible (not married to personnel in pay grades E-1 to E-5, W-1, W-2, O-1, or O-2). This left us with a total analytic sample of 4,454 spouses. In this report, we refer to this subset of the population as "rank-eligible spouses" but note that they are also eligible because of their own civilian status and their service members' active-duty status. It is possible that some of these spouses were not eligible for MyCAA in the 12 months before the survey because they already received the maximum MyCAA funding in previous years, but it was not possible to discern this from the survey items.

Our focus here is on the subset of 4,454 spouses who were eligible for MyCAA based on rank-eligibility, their own civilian status, and their service members' active-duty status. Thus, RAND's findings should not be interpreted as representing overall survey respondents or military spouses more broadly.

Data Analysis

For our analyses, we used the military personnel record data on service members' service, pay grade, and spouse gender and DMDC-provided information on whether the spouse responded to the online version of the survey or the paper version. From the survey, we selected survey items that provided additional demographic, family and military service characteristics, and variables on spouse education, employment, efforts to find work, financial stability, MyCAA program use, number of nights the service member spent away from home in the last three years, health and well-being, and attitudes related to satisfaction with and remaining in the military.

Most of our analyses compare recent MyCAA users to nonusers. To identify users and nonusers, we used reported information on MyCAA use in the last year. In the 2012 ADSS survey, spouses were asked, "Have you used a My Career Advancement

---

1 Similar comparisons of responses on survey items not addressed in this report or for the entire population of 2012 ADSS participants were beyond the scope of this study.

Account (MyCAA) Scholarship in the past 12 months? Mark one." Possible response categories included (1) yes, (2) no, and I was not aware of this resource, or (3) no, but I am aware of this resource. This information is used in subsequent chapters to distinguish between MyCAA users and nonusers. In some analyses, we also limit comparisons to only nonusers who are aware of the program.

This document reports univariate, bivariate, and multivariate analyses for the weighted analytic sample. Where the number (*n*) of respondents is reported, that *n* is unweighted. Most analyses reported are descriptive and used chi-square tests to determine whether differences between MyCAA users and nonusers are statistically significant at the $p < 0.05$ level. More specific details regarding the three types of analyses are provided next.

Recent MyCAA users and nonusers. Using our analytic sample of rank-eligible spouses, described above, we obtained descriptive information on MyCAA use in the last year. Next, we examined weighted percentages of a number of demographic, military, and family characteristics to see whether they differed significantly for MyCAA users and nonusers. We also used multivariate logistic regression models to explore demographic differences among MyCAA users and nonusers, and we report significant differences at the $p < 0.05$ level. Additional detail on these variables is provided in Chapter Three. Finally, we examined nonusers' reported reasons for not using MyCAA in the last year.

Education. To investigate educational goals and reported barriers to pursuing those goals, we once again used our analytic sample of rank-eligible spouses. We obtained descriptive information on the weighted percentage of spouses (overall and by MyCAA use) who report being currently enrolled in school or training. We also examined the educational levels and credentials of spouses who report not being in school but wanting to be to see if current levels of schooling might be related to the decision to go back to school. Finally, we looked directly at reported barriers to meeting schooling desires among spouses who would like to be in school but were not. We also compared these reports for MyCAA users and nonusers to see whether reported differences vary significantly between these groups.

Employment. Drawing again on the analytic sample of rank-eligible spouses, we first obtained descriptive information on the weighted percentage of spouses (overall and by MyCAA use) who were currently working. Among those who were not working and not looking for work, we examined reported reasons for not looking for work. We looked at reported reasons for all spouses in our sample who were not looking for work and also for the subset of spouses who reported that they were not looking for work but wanted to be working. We then turned to job characteristics. We began by examining underemployment—working part-time instead of full-time for those who wished to work full-time. We examined the percentage of MyCAA users and nonusers who work part-time and looked at the main reason reported for pursuing part-time work. Finally, we examined other factors that might point to "successful employment," including whether spouses are currently employed within their area of education or training, a score capturing workplace satisfaction, the time it took to find employment after last PCS move, whether they needed a new license or credential to work at the new duty location, and, if so, the time it took to acquire a new license or credential.

Limitations of Analyses

This study is one of the first to provide a comprehensive look at the barriers keeping military spouses eligible for MyCAA from pursuing and achieving educational and employment goals, yet there are several limitations that should be acknowledged. First, the 2012 ADSS data provide information only about MyCAA use in the past year. There is no information on spouses' history of MyCAA use in these data. Therefore, we do not know how long spouses were using the program or how close they were to completion of the program. When we compared MyCAA users to nonusers, we actually compared recent users to spouses who have not used MyCAA in the last year. However, these nonusers may have used MyCAA sometime in the less recent past. Relatedly, we do not know about people who applied to use the MyCAA program but were rejected.

We also do not have information on why spouses used MyCAA. Users may include spouses who were preparing for a first career, those who were recertifying for a career change, or those who were getting new credentials after a PCS move. The group of users we consider in these analyses may be quite heterogeneous in terms of their reasons for use, but we cannot distinguish different users with these data.

Another limitation is that the 2012 ADSS has information collected only at one point in time (although some questions may be asked about the last month, year, etc.). Given that questions on MyCAA use and characteristics of spouses are asked contemporaneously, we often cannot distinguish cause and effect. For example, when investigating the financial situation of those using MyCAA in the last year, an observed relationship may occur because MyCAA use affects income or because individuals with certain income levels are more likely to use MyCAA. Thus, we cannot speak to the timing of events with these data. On a related note, a fundamental problem in assessing cause and effect also arises from the fact that participation in MyCAA is voluntary rather than randomly assigned across spouses. With voluntary participation, there is always the possibility that any differences in outcomes observed between MyCAA users and nonusers may be in part or wholly a function of unobserved differences (i.e., differences not captured by the survey, such as personal motivation and initiative) between those who do and do not choose to use the program.

It is also worth noting that although the detailed information reported by spouses is part of what makes this dataset ideal for these analyses, self-reports may also be prone to bias, inaccuracies, or recall error. For most of the variables investigated—

which are not sensitive or difficult to recall—we do not expect this to introduce significant bias. As with all surveys, however, it is possible that there is some nonresponse bias for which we cannot account. For example, the survey likely excludes spouses who are uninterested in interacting with the military community or whose service members try to keep them isolated from it, and we do not know whether or how the responses of those spouses might differ from those captured here.

As a concluding note, we caution readers that RAND's analyses of a subset of military spouse responses on the 2012 ADSS do not constitute an evaluation of MyCAA program effectiveness. Additionally, RAND's findings should not be interpreted as representing overall 2012 ADSS survey respondents or military spouses more broadly.

CHAPTER THREE

Use of MyCAA

This chapter explores who is and who is not using the MyCAA scholarship program. First, we review what previous research has documented about the potential need for such a scholarship and the factors that can explain spouse utilization of military programs in general. Next, we report statistics on MyCAA use in the past year for the rank-eligible spouses in our analytic sample, including demographic, family, and military characteristics. We also analyze spouses' reported reasons for nonuse of MyCAA.

Background

An abundance of evidence suggests that military spouses are less likely to be employed than are their civilian counterparts, and that those who are employed earn less (Hosek et al., 2002; Harrell et al., 2004; Lim, Golinelli, and Cho, 2007; Hisnanick and Little, 2014). Some of these differences may be due to basic differences in the background and demographic characteristics between military and civilian spouses. For instance, military husbands and wives are much more likely to relocate than their civilian counterparts (Lim, Golinelli, and Cho, 2007). Military spouses are also younger, less likely to be white, and more likely to have young children at home (Lim, Golinelli, and Cho, 2007). The unpredictable, heavy, and rigid work schedules of active-duty service members is another factor that may explain differences in the employment and earnings of spouses (Hosek et al., 2002; Castaneda and Harrell, 2008).

Postsecondary education, including the pursuit of occupational licenses and certificates, is one way to narrow the earnings and educational gap between military and civilian families. Additional education could improve a military spouse's job prospects by providing the skills and credentials needed to obtain certain jobs. In previous research, spouses expressed a preference for more education rather than pursuit of a credential, but more research is needed to understand why (White House, 2011, p. 16; Needels and Zaveri, 2009). Although many studies have focused on understanding military spouses' employment preferences, challenges, and earnings, little research sheds light on spouses' educational preferences, goals, and challenges to reaching those goals.

Not all military spouses who could benefit from educational and employment assistance programs take advantage of them, however. The literature provides several explanations as to why this may be the case. For example, spouses have reported problems accessing military spouse employment programs, including inconsistency in the information provided to spouses across different sources, lack of awareness of resources available, and difficulty interfacing with websites and portals (U.S. Government Accountability Office [GAO], 2012b, pp. 10–11). A 2009 survey of Air Force spouses found that among respondents experiencing employment problems, "around 30 percent reported each of the following about Air Force employment assistance programs: lack of information, inconvenient access, and inability to address their problems" (Miller et al., 2011a, p. 25). Additionally, redundancy of programs may result in inefficient utilization.

Other factors relating to demographic and family circumstances may explain the likelihood of spouses using educational and employment military programs. Issues with child care, for instance, have been cited as a significant factor related to the employment of wives of military service members (Castaneda and Harrell, 2008; Lim, Golinelli, and Cho, 2007; Schwartz, Wood and Griffith, 1991; Zellman et al., 2009). Difficulty obtaining child care can be especially pronounced in the military setting, where service members frequently have inflexible work hours, frequent moves, and absent spouses as a result of deployments. Military spouses often have to operate as single parents when their service members are away on temporary assignments or deployments or working in positions with extremely long hours (e.g., drill instructor). In previous research, military spouses have reported service member absence as a work-related challenge, particularly when the couple has children (Castaneda and Harrell, 2008). Working spouses of military members report more unmet child care needs than single military members, because single parents in the military are given priority for child care (Zellman et al., 2009, p. 448).

Where military families live may also be related to use and awareness of assistance programs. For example, in the past, junior enlisted Army spouses reported isolation and detachment from the military community when they lived off base (Harrell, 2000). There is some evidence of demographic differences in housing assignment. According to a telephone sample of about 800 Air Force spouses, spouses living on base were younger, slightly less educated, less likely to be in a dual-military couple, less likely to be married to an officer, and less likely to be working full-time than their counterparts off base (Miller et al., 2011a, p. 12). Given that junior enlisted personnel have lower rates of military pay than other personnel, financial constraints may be more likely to affect where they live, and, consequently, their awareness of education and employment opportunities both on and off base. Living off base increases the likelihood of spouse employment but can also restrict access to on-base resources (Miller et al., 2011a).

Research has just begun to separately consider the needs of the smaller population of male civilian spouses (Hisnanick and Little, 2014; Lim, Golinelli, and Cho, 2007). Historically, the U.S. military population has been nearly all or mostly male, and thus programs aimed at military spouses were very practically aimed at women (Segal, 1986). Still, features of military life appear to affect male spouses' careers as well, as they, too, experience a wage gap relative to the husbands of civilian women (Cooney, De Angelis, and Segal, 2011; Hisnanick and Little, 2014). In addition, one study found that whereas civilian husbands married to military women had better employment outcomes than did civilian wives married to military men, husbands were less satisfied with their outcomes than were wives. Specifically, the likelihood of a civilian wife being dissatisfied with employment outcomes is about one-third lower than the likelihood of a civilian husband being dissatisfied (Cooney, De Angelis, and Segal, 2011).

In sum, many possible demographic and military service factors should be considered when exploring why some spouses use educational assistance programs and others do not.

Who Uses MyCAA?

The 2012 ADSS asked respondents if they had used a MyCAA scholarship in the past 12 months. We first determined the percentage of eligible spouses who used MyCAA in the last year and then compared the demographic, service, family, and other general characteristics of MyCAA users and nonusers.

About One in Five Rank-Eligible Spouses Used a MyCAA Scholarship in the Last Year

Table 3.1 depicts recent MyCAA use based on self-reported use of this resource in the last year. Among rank-eligible spouses, 18 percent had used MyCAA in the previous year. Of those who did not use MyCAA, 54 percent indicated they were unaware of the program.

Table 3.1
Reported Use of a MyCAA Scholarship in the Last 12 Months (n = 4,454)

	Rank-Eligible Spouses (%)
Users	18
Nonusers	82
Aware of MyCAA	46
Unaware of MyCAA	54

We first examine the differences between users and nonusers and then focus on the nonusers to explore differences between those who were aware of MyCAA and those who were not.

Demographic Factors Differed Considerably Between Users and Nonusers

To assess the factors that are correlated with MyCAA use, we examined a number of characteristics of spouses, service members, and their families more generally. Tables 3.2–3.3 and 3.5–3.6 report descriptive statistics for a series of spousal characteristics grouped by whether rank-eligible spouses used MyCAA in the last year. Group differences were tested with chi^2 tests to determine whether factors differed significantly for users and nonusers.

Characteristics of Spouses

As shown in Table 3.2, nearly all the demographic factors considered here differed significantly among MyCAA users and nonusers. MyCAA users were younger than nonusers (44 percent of users are under age 26 compared to 37 percent of nonusers), more likely than nonusers to be female (97 percent of users compared to 93 percent of nonusers), less likely to be white (59 percent of users compared to 66 percent of nonusers), and much more likely to fall into the "some college" category (indicating that they have some college but do not have either an associate's degree or bachelor's degree (57 percent of users compared to 37 percent of nonusers). MyCAA users were also significantly more likely than nonusers to have earned a vocational or technical certificate after high school (29 percent of users compared to 18 percent nonusers). MyCAA users were also more likely to have no income at all (38 percent of users compared to 32 percent of nonusers).

Characteristics of Service Members

Table 3.3 shows service members' characteristics for users and nonusers. We found significant differences between users and nonusers in their husbands' and wives' service and pay grade, as shown in Table 3.3. MyCAA users were more likely than nonusers to be married to service members in the Army (50 percent compared to 44 percent) and also more likely to be married to service members in junior enlisted pay grades of E-1 to E-4 (62 percent compared to 55 percent).

Spouses of Junior Enlisted Personnel Were Most Likely to Lack a College Degree

As noted in the previous chapter, eligibility for MyCAA is limited to spouses of service members in certain pay grades and for particular courses of study or testing. Table 3.4 shows the education levels of spouses in our analytic sample, by the pay grades of their marital partners. Spouses of higher-ranking service members were significantly more likely to have a college degree than spouses whose service members are lower in the hierarchy. About 70 percent of the spouses of junior officers O-1 to O-2 and 50 per-

Table 3.2
Demographic Characteristics of Spouses, by MyCAA Use

		MyCAA Users (%)	MyCAA Nonusers (%)
Age (years)[a]	< 26	44	37
	26–30	37	35
	31–35	11	16
	36–40	5	7
	> 40	4	4
Gender[a]	Female	97	93
	Male	3	7
Race-ethnicity[a]	White	59	66
	Nonwhite	41	34
Education[a]	Less than 12 years of school (no diploma)	1	3
	High school graduate (diploma or equivalent)	10	15
	Some college (no degree)	57	37
	Associate's degree (e.g., AA, AS)	16	15
	Bachelor's degree (e.g., BA, AB, BS)	13	24
	Master's degree or more	3	7
Vocational certificate[a]	Received vocational or technical diploma	29	18
	Did not receive	71	82
Own monthly income[a]	None	38	32
	More than 0 but ≤ $1,000	14	14
	More than $1,000–≤ $2,000	11	13
	More than $2,000–≤ $3,000	8	12
	More than $3,000–≤ $5,000	12	12
	More than $5,000–≤ $7,000	2	3
	More than $7,000	1	2
	Missing income information	13	11

NOTE: Sample sizes vary slightly because of missing data.

[a] MyCAA users differ significantly from MyCAA nonusers at $p < 0.05$ based on chi^2 test.

cent of the spouses of warrant officers W-1 to W-2 already had a bachelor's or graduate degree.[1] Spouses of junior enlisted personnel (E-1 to E-4) were most likely to have some college but not yet a degree (45 percent). Differences in the distribution of education by pay grade may explain differences reported above in MyCAA use by pay grade.

[1] Note that a bachelor's degree is required to become a military officer but not a warrant officer, NCO, or junior enlisted member of the military.

Table 3.3
Spouse's Service Member Characteristics, by MyCAA Use

		MyCAA Users (%)	MyCAA Nonusers (%)
Service[a]	Army	50	44
	Navy	15	20
	Marine Corps	17	15
	Air Force	17	22
Pay grade[a]	Junior enlisted (E-1 to E-4)	62	55
	Noncommissioned officer (NCO) (E-5)	33	37
	Warrant officer (W-1 to W-2)	2	3
	Junior officer (O-1 to O-2)	3	6
Nights in the last 36 months service member was away from home	0	5	5
	1–59	15	17
	60–179	17	17
	180–269	15	14
	270–449	25	25
	450–1,095	19	19
	Missing	3	3

NOTE: Sample sizes vary slightly because of missing data.
[a] MyCAA users differ significantly from MyCAA nonusers at $p < 0.05$ based on chi^2 test.

Family Characteristics

Table 3.5 depicts a variety of characteristics of the family, by use status. Just over one-third of both users and nonusers had been married less than three years, but MyCAA users were more likely than nonusers to have been married between 3 and 6 years (42 percent of users compared to 35 percent of nonusers). As shown in this table, MyCAA users were more likely than nonusers to be living on base (36 percent compared to 30 percent) and less likely than nonusers to be living in civilian housing (56 percent compared to 51 percent).

Health, Financial Well-Being, and Satisfaction with the Military

Finally, there were also differences in users' and nonusers' financial well-being and satisfaction with the military, as shown in Table 3.6. Levels of stress and physical and mental health did not differ significantly between the two groups. MyCAA users were worse off financially than nonusers. They were more likely to have at least one bad

Table 3.4
Spouse Education, by Service Member's Pay Grade

	Junior Enlisted (E-1 to E-4) (%)	NCO (E-5) (%)	Warrant Officer (W-1 to W-2) (%)	Officer (O-1 to O-2) (%)	Total (%)
Less than 12 years of school (no diploma)	3	2	<1	<1	2
High school graduate (diploma or equivalent)	16	12	6	3	14
Some college (no degree)	45	37	26	17	40
Associate's degree (e.g., AA, AS)	14	18	17	9	15
Bachelor's degree (e.g., BA, AB, BS)	18	24	38	51	22
Master's degree or more	4	7	14	20	6

Table 3.5
Family Characteristics, by MyCAA Use

		MyCAA Users (%)	MyCAA Nonusers (%)
Years married[a]	<3	34	36
	3–6	42	35
	6–10	16	19
	10–15	6	7
	15+	2	2
Children	None living at home	39	38
	Age <6 years	34	37
	Age 6–13 years	22	20
	Age 14–18 years	5	5
Enrolled in the Exceptional Family Member Program (EFMP)	Yes	10	10
	No (or not applicable)	90	90
Housing[a]	On base	36	30
	Military off base	9	9
	Civilian housing	56	61

NOTE: Sample sizes vary slightly because of missing data.

[a] MyCAA users differ significantly from MyCAA nonusers at $p < 0.05$ based on chi^2 test.

Table 3.6
Health, Financial Well-Being, and Satisfaction with the Military, by MyCAA Use

		MyCAA Users (%)	MyCAA Nonusers (%)
More stress than usual in personal life		58	54
Above average mental health		56	59
Above average physical health		60	56
Financial condition	Very comfortable/able to make ends meet	46	51
	Occasionally have some difficulty	32	29
	Difficulty making ends meet/in over our heads	22	19
At least one bad financial outcome[a]		35	30
Savings ≥ $500[a]		45	54
Satisfied with the military		57	58
Favor staying in the military		66	66

NOTE: Sample sizes vary slightly because of missing data.

[a] MyCAA users differ significantly from MyCAA nonusers at $p < 0.05$ based on chi^2 test.

financial outcome (35 percent compared to 30 percent) and less likely than nonusers to have at least $500 in savings (45 percent compared to 54 percent). As a reminder, we cannot assume causality in these analyses: A difficult financial situation might be the reason these spouses use the MyCAA program in the first place, or it could be a short-term consequence of being in school or training. There were no differences between the two groups in terms of satisfaction with the military or preferring that their marital partner remain in military service.

Demographic Factors Also Differed Considerably Between Aware and Unaware Nonusers

We conducted several additional analyses using the survey responses of those who had not used MyCAA in the 12 months before the survey to look for any statistically significant differences ($p < 0.05$) between the characteristics of those who were aware of the scholarships and those who were not. A description of the survey items capturing the characteristics of spouses, service members, and their families is discussed in more detail in Appendix A.

The largest differences we detected suggest that opportunity for exposure to information about MyCAA plays a role in spouse awareness of the scholarships. Newly married spouses were far less likely to be aware of MyCAA: 43 percent of spouses who were unaware of MyCAA had been married for less than three years compared to 28 percent of spouses who were aware of it. Nonusers unaware of MyCAA were more likely than aware nonusers to live in civilian housing, more likely to be married to junior enlisted service members (E-1 to E-4), and less likely to be married to NCOs (E-5) (6–7 percentage point differences). Additionally, spouses who were aware of the program were more likely to be women than spouses who were unaware (96 percent compared to 90 percent). More details about these comparisons are provided in Tables B.1–B.3.

Factors Associated with MyCAA Use Included Being Female, Nonwhite, Some College but No Degree, No Income, No Children Under Age Six, and Living on Base

We included the same factors from Tables 3.2–3.6 in a multivariate logistic regression model of MyCAA use. Specifically, we looked at the association between users or nonusers and a variety of factors, including age, gender, race-ethnicity, education, vocational or technical diploma, years married, children, service, pay grade, enrolled in the EFMP, housing, nights in the last 36 months the service member was away from home, more stress than usual in personal life, below average mental health,[2] below average physical health, own monthly income, financial condition, at least one bad outcome, savings > $500, satisfied with the military, favor staying in the military, and completed mail survey. Given that so many characteristics differed for these groups, we modeled all variables simultaneously so that we could see which factors stood out even after controlling for all others. The results of this analysis are reported in detail in Table C.1. We highlight all the statistically significant findings below:

- Net of all factors, women remained much more likely than men to have used MyCAA in the last year. More specifically, women were 2.58 times more likely than men to be MyCAA users.
- In addition, whites were 71 percent less likely than nonwhites to use MyCAA.
- In terms of educational attainment, those with some college but no degree were significantly more likely to be MyCAA users than were those with any other level of education. Those without a high school diploma or equivalent at the time of the survey were least likely to have used MyCAA in the previous year. Since the survey asked about past use of MyCAA, it is perhaps not surprising that MyCAA users at the time of the survey would be least likely to be lacking a high school diploma or equivalent. If a spouse had not already completed that education level, MyCAA could have supported it as a first step within an approved plan for a license, certificate, or associate's degree.
- MyCAA users were significantly more likely than MyCAA nonusers to hold vocational or technical degrees.

---

2 Mean mental and physical health measures for this population were constructed on the analytic sample of rank-eligible spouses. These measures were constructed from the PHQ-4 (mental health) and general health perceptions subscale on the Short-Form Health Survey (physical health). Measures are described in more detail in Appendix A.

- Spouses married for a few years (3–6) were about one-and-one-half times more likely to be MyCAA users than newlyweds married less than three years.
- Although not significant in the univariate models, in the fully adjusted models, spouses with children age 6 or younger were significantly less likely to be MyCAA users than are spouses with no children.
- Spouses with partners in some services were significantly more likely than others to have used MyCAA in the last year. Spouses with husbands or wives in the Navy or Air Force were disproportionately less likely than those in the Army to be MyCAA users. This analysis already accounts for the fact that there are more Army spouses overall.
- Spouses living on base were significantly more likely than those living in civilian housing to use MyCAA.
- Finally, although financial factors were related to MyCAA use in the univariate analyses, once we adjusted for demographic characteristics, these factors no longer significantly differed by MyCAA use. The one exception is income. Spouses with no income were more likely to have used MyCAA in the last year, although we cannot determine from these data whether this is a cause of using the program (e.g., MyCAA use because of low income or poor work prospects) or a consequence (e.g., not working because of being back in school). We explore the relationship between MyCAA use and employment outcomes in more detail in Chapter Five.

Table C.1 also reports the same analyses, this time for a subsample of the original analysis, that is, nonusers who are aware of MyCAA. For the most part, results are very similar. One noteworthy difference is that, on excluding MyCAA nonusers who stated on the survey that they were unaware of the scholarships, we no longer see significant differences by service. This may suggest that these differences in use are due to awareness of MyCAA. To check whether this is the case, we ran an additional set of analyses (not shown) looking only at MyCAA nonusers and comparing aware nonusers to unaware nonusers. We found that spouses of service members in the Navy and Air Force were significantly less likely than those in the Army to report being aware of the scholarships (Marines were less likely but not statistically significantly different from those in the Army). This was true even after controlling for all other factors. Housing also mattered less when we considered only aware nonusers, suggesting, too, that living on base may foster awareness of military programs.[3]

[3] In sensitivity analyses not displayed here, we also considered other constructions of the model variables, including a seven-category and logged version of spousal income and linear versions of mental and physical health. For the most part, results were substantively similar to those presented here. For income, however, results were not significant when we considered the seven-category version of the variable. Income differences result mainly because MyCAA users are less likely than nonusers to have any income at all.

What Are the Reasons for Not Using MyCAA?

Those spouses who reported that they did not use MyCAA in the last year but indicated that they are aware of the resource were asked a follow-up question about the reason for nonuse. Specifically, spouses were asked, "What is the main reason you did not use a My Career Advancement Account (MyCAA) Scholarship in the past 12 months? Mark one." Responses included (1) I am not eligible because of my husband's/wife's rank, (2) I am not eligible because my level of education enrollment does not qualify, (3) I have limited time for additional education/training because of family/personal obligations, (4) I am not interested in additional training/education, and (5) I do not feel that additional education and training are important for my career.

Some Spouses Who Appear to Be Rank-Eligible Did Not Perceive That They Were

We first examined whether perceived rank-ineligibility reported as a primary reason for nonuse matched rank-ineligibility based on the rank documented in service members' administrative records. Table 3.7 shows the results of this analysis. About 7 percent (n = 119) of rank-eligible spouses who knew about MyCAA but did not use it in the previous year thought that their service member's rank was ineligible for MyCAA when in fact they were eligible based on service records. Most of those spouses were married to E-5s. In fact, 63 percent of those who reported ineligibility as their primary

Table 3.7
Perceived Rank Ineligibility Among Nonusers Who Were Aware of MyCAA, by Service Member Pay Grade

	Rank Ineligibility Is Primary Reason for Nonuse (n = 119) (%)	Rank Ineligibility Not Given as Primary Reason for Nonuse (n = 1,505) (%)
E-1	0	1
E-2	0	2
E-3	4	15
E-4	9	37
E-5	63	38
W-1	4	1
W-2	8	1
O-1	2	3
O-2	10	3

NOTE: Only spouses who reported that they did not use MyCAA but that they were aware of it were asked this question.

reason for nonuse were married to E-5s, compared to 38 percent of the rest of the population. This suggests that one reason for nonuse may be a mismatch between perceptions of rank-eligibility for these scholarships and actual eligibility. It is possible that these perceptions are correct if their service members are about to be promoted out of eligibility, but there could also be a lack of understanding about the eligibility criteria among some spouses. In analyses not reported here, we also examined ineligibility by service but did not find substantive differences by service.

Perceptions of Educational Ineligibility Rose with Increased Education Level

Twenty-eight percent of rank-eligible spouses who knew about MyCAA but did not use it in the previous year thought that their level of educational enrollment made them ineligible for MyCAA. Although education alone does not make a spouse ineligible for the MyCAA program, we looked at whether perceived educational ineligibility reported as a primary reason for nonuse was more commonly reported by spouses with higher education levels. Table 3.8 shows the results of this analysis. By comparing the column on the left to that on the right, we can see where the education levels of spouses who gave educational ineligibility as a primary reason for nonuse (left) differ from those of the spouses who did not offer this reason (right). As expected, spouses who reported this form of ineligibility had more advanced educational credentials than spouses who did not report this as a primary reason for not using the MyCAA program in the last year. In fact, 56 percent of spouses who reported that the primary reason for ineligibility was due to educational requirements had bachelor's degrees or higher compared to only a quarter of those spouses who did not report this as a primary reason.

Table 3.8
Perceived Educational Ineligibility Among Nonusers Who Were Aware of MyCAA, by Education Level

	Educational Ineligibility Given as Primary Reason for Nonuse (n = 453) (%)	Educational Ineligibility Not Given as Primary Reason for Nonuse (n = 1,171) (%)
Less than 12 years of school (no diploma)	3	2
High school graduate (diploma or equivalent)	1	19
Some college (no degree)	21	41
Associate's degree (e.g., AA, AS)	19	14
Bachelor's degree (e.g., BA, AB, BS)	43	20
Master's degree or more	13	5

NOTES: Education level alone does not make one ineligible for MyCAA. Only spouses who reported that they did not recently use MyCAA but that they were aware of it were asked this question.

As noted above, although MyCAA scholarships do not fund the pursuit of bachelor's or graduate degrees, spouses at any education level are eligible for funds to support professional licensing or certification in approved occupations (e.g., a state bar exam, a commercial pilot's license, pre-license courses, and exam for a real estate license). A spouse with a Ph.D. in philosophy is still eligible for MyCAA funding to sponsor attainment of a portable career-focused associate's degree, license, or certificate.

The Primary Reasons for Nonuse Among Spouses Aware of MyCAA Were Time Constraints and Perceptions of Ineligibility

We show the main reason spouses reported for not using MyCAA in more depth in Table 3.9. The most frequently reported reason for not using MyCAA, among those aware of the scholarships, was limited time. The majority of spouses reported this as the main reason for not using MyCAA. This is consistent with civilian employer educational assistance programs, which also report time as the biggest barrier to program utilization (Lucal, 2012). As discussed above, we also found that 28 percent of respondents in our sample believed that they were ineligible for MyCAA because of their level of education. Again, technically, no spouses are ineligible for MyCAA because of their level of education.

If we exclude spouses who reported ineligibility as their primary reason for not using MyCAA, nearly 80 percent of all spouses in our sample who were aware of MyCAA reported time as the primary barrier preventing them from using this program. Only 15 percent reported that their primary reason for not using MyCAA was that they were not interested in obtaining more schooling or training, and only 5 percent reported that the main reason they did not use MyCAA was because they did not

Table 3.9
Frequencies of Main Reason Reported for Not Using MyCAA, Among Spouses Who Were Aware of It

	All Rank-Eligible Nonusers Aware of MyCAA (n = 1,624) (%)	Only Those Who Perceive Themselves As Eligible (n = 1,052) (%)
Ineligible because of my husband's/wife's rank	7	—
Ineligible because my level of education does not qualify	28	—
I have limited time for additional education/training because of family/personal obligations	52	79
Not interested in additional education or training	10	15
Do not feel that additional education or training is important for my career	4	5

NOTE: Only spouses who reported that they did not use MyCAA but that they were aware of it were asked this question.

feel it was important for their career. It is possible that multiple reasons guided spouses' decisions not to take advantage of this program, but it is noteworthy that lack of time was by far the most commonly reported barrier to program utilization.

Chapter Summary

There are many reasons posed in the literature for why spouses may not use a program such as MyCAA, including lack of demand, lack of information or inconsistent or inaccurate information, and demographic and family circumstances. We examined the use of MyCAA among eligible spouses, looked at the differences between users and nonusers, and consider the reasons that eligible spouses do not use these scholarships.

In our sample of rank-eligible spouses, 18 percent of eligible spouses had used MyCAA in the previous year. Of those who did not use MyCAA, 54 percent reported that they were unaware of it. The lack of program awareness among spouses is consistent with other work examining why military spouses do not use available services (GAO, 2012b, pp. 10–11). We also point to another type of awareness that may be lacking—awareness of rank eligibility. Insufficient time is the most common factor reported for not using MyCAA in the last year.

Using multivariate models that allow us to examine the associations between MyCAA use and a variety of demographic variables while accounting for other factors, we also found evidence of statistically significant differences between recent MyCAA users and nonusers. MyCAA users were younger than nonusers, more likely to be female, less likely to be white, much more likely to have some college but no degree, and more likely to have earned a vocational or technical certificate. Compared to nonusers, MyCAA users had lower reported income and less savings, were more likely to live in military on-base housing, and were more likely to be married to enlisted personnel in the E-1 to E-4 pay grades. Users were less likely than nonusers to have children age 6 or under living at home and were disproportionately less likely to be married to Navy or Air Force personnel. Although some of these differences are likely related to spouse need or program requirements, others may point to a lack of awareness or access to these programs among certain populations.

CHAPTER FOUR

Higher Education Preferences, Experiences, and Barriers

Education can provide spouses with more opportunities for promotion at their current job and might also directly improve a spouse's job prospects. Higher levels of education can potentially compensate for the absence of continuous work histories by signaling to employers that spouses nonetheless have the skills needed to perform the job. MyCAA is designed to help eligible spouses establish and reach educational goals for associate's degrees, occupational licenses, and certificates in high-growth, high-demand career fields. In this chapter, we consider how well MyCAA is helping eligible spouses establish and attain their educational goals. Specifically, we examine (1) the percentage of spouses who are in school and, for those who are not, whether they want to be, and (2) for those spouses who are not in school but want to be, the reasons why they are not pursuing additional training and schooling.

Are MyCAA Users More Likely to Attend Schooling or Training?

Recent MyCAA Users Were More Likely Than Nonusers to Be in School

As Figure 4.1 depicts, our analysis found that 28 percent of rank-eligible spouses were enrolled in school or training at the time of the survey and many more would like to be: 46 percent. Recent MyCAA users were more likely than nonusers to be in school (62 percent compared to 20 percent). However, 25 percent of recent MyCAA users compared to 51 percent of rank-eligible MyCAA nonusers wanted to be in school at the time of the survey but were not. Thus, a fair number of recent MyCAA users—25 percent—wanted to be in school but were not. The difference between users and nonusers was statistically significant ($p < 0.001$).

The Education Levels for Spouses Who Wanted to Be in School but Were Not Differed for Recent MyCAA Users and Nonusers

Spouses might not be in school or training at the time of the survey for many possible reasons. First, we looked at reported education levels and credentials of spouses who were not in school but wanted to be by whether they were MyCAA users. These results

Figure 4.1
Percentage of Survey Respondents Enrolled in School or Training

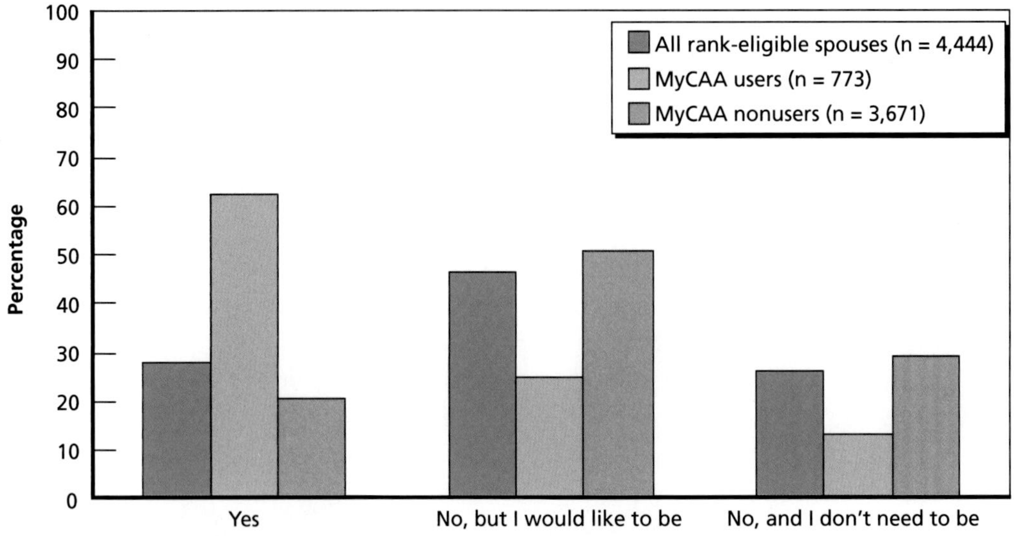

RAND *RR784-4.1*

are shown in Table 4.1, first for all rank-eligible spouses, then distinguishing between MyCAA users and nonusers.

Fifty-five percent of spouses who were not in school but would like to be reported having completed some college or an associate's degree—groups that are likely eligible for financial assistance through MyCAA. Not surprisingly, many more MyCAA users than nonusers fell into these two categories (73 percent compared to 53 percent), and educational levels differed significantly ($p < 0.05$) for MyCAA users and nonusers. Of spouses who wanted to be in school, 19 percent held a high school diploma or equivalent and 20 percent had a bachelor's degree. Only 11 percent of MyCAA users fell into the high school diploma category, and only 13 percent of MyCAA users had a bachelor's degree. Finally, about 20 percent of spouses who wanted to be in school but were not already held a vocational degree. However, recent MyCAA users who wanted to be in school but were not were much more likely than nonusers to have such a diploma (40 percent compared to 20 percent). These results also suggest that for higher levels of schooling, spouses with the desire for additional education or training may be left with unmet needs because of the eligibility requirements of MyCAA.

The Cost of Education Was the Key Factor Preventing Spouses from Attending School or Training

The 2012 ADSS study also provides direct information on the reasons spouses report for not being in school. Spouses who reported wanting to be in school but were not were asked to select from a list of eight possible reasons for not being in school. Spouses

Table 4.1
Education Levels Among Those Who Would Like to Be in School or Training but Who Are Not, Overall and by MyCAA Use

		All Rank-Eligible Spouses (n = 2,009) (%)	MyCAA Users (n = 185) (%)	MyCAA Nonusers (n = 1,824) (%)
Education level[a]	Less than 12 years of school (no diploma)	3	1	3
	High school graduate (diploma or equivalent)	19	11	20
	Some college (no degree)	42	58	40
	Associate's degree (e.g., AA, AS)	13	15	13
	Bachelor's degree (e.g., BA, AB, BS)	20	13	20
	Master's degree or more	3	3	4
Technical/vocational diploma[a]	Has vocational diploma	22	40	20
	Does not have vocational diploma	78	60	80

[a] MyCAA users and nonusers differ significantly at $p < 0.05$.

could select multiple reasons. Table 4.2 shows this list of reasons for the sample of spouses overall and broken down by MyCAA users and nonusers. Both MyCAA users and nonusers most commonly cited the cost of education as a reason, with 82 percent of rank-eligible respondents listing this reason. This finding demonstrates that the MyCAA program is targeting a current need by providing a way for rank-eligible spouses to address some of the costs of education, although, of course, the program does not offer unlimited resources for postsecondary education.

A number of possible scenarios could explain why recent MyCAA users might not be in school even if they want to be. For example, spouses who have already obtained an associate's degree may be interested in earning a bachelor's degree. Additionally, some career fields offer increasing levels of certification to demonstrate advanced skills. Some spouses may wish to pursue multiple occupational certificates in their career field to increase their abilities and job or promotion opportunities (e.g., certificates in medical assistance and phlebotomy or in web design and web development) but at some point hit their MyCAA scholarship limit. Indeed, MyCAA program data (not shown here) reveal that some spouses pursued more than one educational objective.

Other Reasons for Not Pursuing Further Education and Training Included Family Responsibilities and Child Care Costs

Family responsibilities and the expense of child care were also cited by more than half of MyCAA users and nonusers as reasons for not pursuing additional education or

Table 4.2
Factors That Prevent Spouses Who Would Like to Attend School or Training from Doing So, by MyCAA Use

	% Reporting "Yes"		
	All Rank-Eligible Spouses (n = 2,009)	MyCAA Users (n = 185)	MyCAA Nonusers (n = 1,824)
Cost of education	82	76[a]	82[a]
Family responsibilities	68	65	69
Expense of child care	53	54	53
Hours/location not convenient	45	40	45
Conflicts with work schedule	37	36	37
Sponsor's deployment makes it difficult	36	36	36
Other	29	28	29
Move too often	25	27	25
Transportation problems	23	27	22

[a] MyCAA users and nonusers differ significantly at $p < 0.05$.

training. Deployment was also cited by just over one-third of spouses as a reason for not pursuing education.

Moving too often and transportation issues were the least-cited reasons at 25 percent and 23 percent, respectively, for the overall sample of spouses. Still, these are issues that can potentially be addressed, at least for some professions, through online education. We return to this possibility in our discussion of recommendations in Chapter Six.

MyCAA users and nonusers who wanted to be in school mostly reported similar barriers keeping them from pursuing their education. The cost of schooling was the only reason that differed significantly for MyCAA users and nonusers, with MyCAA users being less likely to report cost as an issue. This is very likely because users of MyCAA have already at least begun the process of receiving financial assistance through the MyCAA program. None of the other reasons differed significantly for MyCAA users compared to nonusers.

Access Issues and Features of Military Life Prevented Some Spouses from Attending School or Training

A number of barriers involved access to school or training. Nearly half (45 percent) of spouses who were not in school but would like to be cited inconvenient school hours or location, 37 percent reported conflicts with work schedules, and 23 percent experienced problems with transportation. For 36 percent of respondents, service member deployments made it difficult to attend school or training, and for 25 percent, frequent moves were the problem. In each of these categories, there were no significant differences in the responses of spouses who recently used a MyCAA scholarship and those who did not.

Chapter Summary

Military spouses described facing many barriers to attaining their educational goals. Our findings corroborated a previous qualitative study, which suggested that the costs of education and child care needs prevent military spouses from pursuing additional schooling (Harrell, 2000). Although educational costs were a bigger concern for MyCAA nonusers, family responsibilities and the cost of child care were educational hurdles similarly shared by both MyCAA users and nonusers. Nationally, the costs of education and child care are common impediments to achieving college education (Miller, Gault, and Thorman, 2011; Perna and Jones, 2013).

A recent GAO investigation found that lack of awareness of DoD-subsidized child care and demand that exceeds availability are the two main barriers preventing spouses from taking advantage of this benefit (2012a). DoD has taken steps to address these problems, including increased outreach and education efforts, the construction of additional on-installation facilities, and initiatives to expand eligible off-installation providers (GAO, 2012a). Nonworking spouses enrolled in accredited postsecondary institutions hold higher priority than other nonworking spouses for allocation of limited space, DoD-subsidized, full-time child care, but they can elevate their priority status by simultaneously being employed (top priority, or Priority 1) or seeking work (Priority 2) (U.S. Department of Defense, 2014). Fees are income-based and may be adjusted case by case to address financial hardships (U.S. Department of Defense, 2014). Other hurdles to pursuing education reported in the survey included access issues related to school location and transportation, inconvenient hours and conflicts with work schedules, and features of military life, such as deployments and frequent moves.

There were large differences between MyCAA users and nonusers in terms of the likelihood of being in school or training at the time of the survey and in wanting to be in school or training. Recent MyCAA users were more likely than nonusers to be in school. In addition, significantly fewer recent MyCAA users than nonusers reported that they would like to be in school but were not. A fairly large percentage of spouses with a high school diploma or equivalent or a bachelor's degree would have liked to be in school but were not. In addition, about 20 percent of spouses who wanted to be in school but were not already hold a vocational degree. This may suggest that, for certain levels of schooling, spouses with the desire for additional education or training may be left with unmet needs because of the eligibility requirements of MyCAA.

CHAPTER FIVE

Employment Preferences, Experiences, and Barriers

MyCAA is designed to help military spouses obtain occupational licenses and certificates and some types of education and training that will serve them well in the job market and reduce unemployment and underemployment among this population. Thus, it is important to examine the extent to which recent MyCAA users and nonusers are working, why they may not be working, and whether they are successfully employed in jobs that are a good match for their training and are satisfying, for instance. More specifically, in this chapter, we summarize our analyses investigating (1) the extent to which spouses are working and looking for work at the time of the survey, (2) reasons for not working or working only part-time, and (3) the job characteristics of employed spouses. In studying these questions, we pay particular attention to differences between spouses in our subset of 2012 ADSS respondents who used MyCAA in the past year and spouses who did not.

How Many Spouses Are Working?

About 40 Percent of Spouses Were Working

Among the overall subset of respondents in RAND's analyses, 85 percent reported wanting to work, and 64 percent reported needing to work. Figure 5.1 shows the percentage of spouses who were working, both overall and broken down by whether they had used MyCAA in the previous 12 months. Thirty-nine percent of all rank-eligible spouses and 34 percent of MyCAA users reported that they worked for pay or profit in the week before the survey. The career fields of the employed spouses in our sample included health care and health services (17 percent), retail and customer service (15 percent), education (11 percent), administrative services (9 percent), child care and child development (6 percent), financial services (6 percent), recreation and hospitality (6 percent), skill trades (4 percent), communications and marketing (3 percent), animal services (2 percent), and information technology (2 percent), and 20 percent were working in other career fields not listed on the survey.

Figure 5.1
Percentage of Spouses Working for Pay or Profit in the Last Week, by MyCAA Use

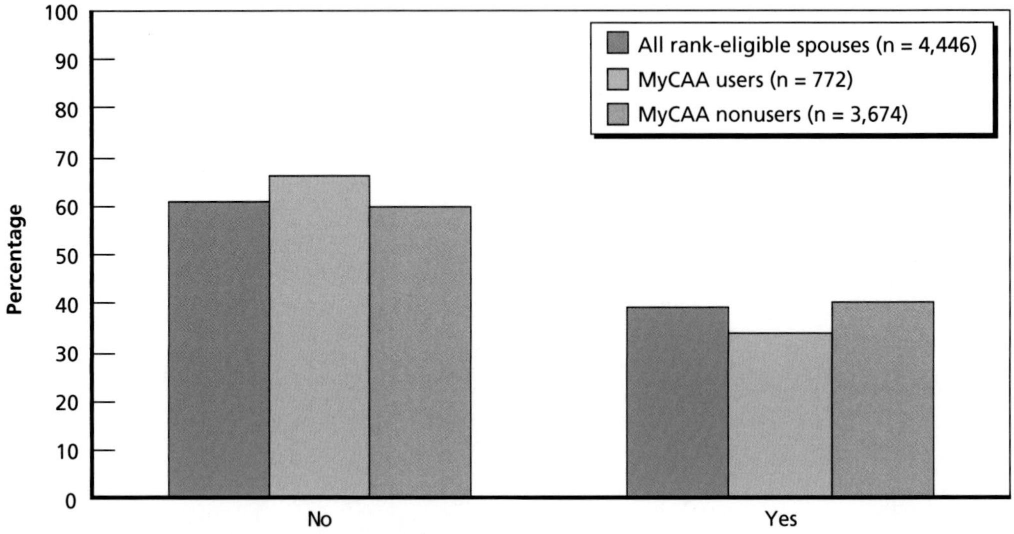

NOTE: MyCAA users differ significantly from nonusers, $p < 0.05$.
RAND *RR784-5.1*

About One-Third of the Unemployed Were Looking for Work

Spouses who reported that they were not working were asked whether they had looked for work in the last month.[1] As Figure 5.2 shows, over one-third of rank-eligible spouses indicated that they were not working and looking for work.

The Vast Majority of Spouses Who Were Not Looking for Work Wanted to Work

In addition to understanding whether spouses were working or looking for work, we were also interested in whether spouses who were not looking for work wanted to be working. Regardless of whether they were currently working, spouses were asked whether they wanted to work (yes or no).

Figure 5.3 shows responses for the subset of rank-eligible spouses who were not looking for work. As shown, the vast majority of spouses not looking for work said that they wanted to work (70 percent). Spouses were also asked whether they needed to work. Thirty-seven percent of spouses who were not looking for work reported that they needed to work (not shown).

---

[1] A small subset of web respondents (348 rank-eligible spouses) were asked whether they did anything to "find work" and what they did to find work, rather than being asked whether they were "looking for work." In sensitivity analyses, we reran Figures 5.2 and 5.3 with the inclusion of those spouses who reported doing something to find work. Results were nearly identical to those reported. This subset of spouses was not asked about reasons for looking for work reported in Table 5.1.

Figure 5.2
Percentage of Spouses Looking for Work in the Last Four Weeks, Among Those Who Are Not Working, by MyCAA Use

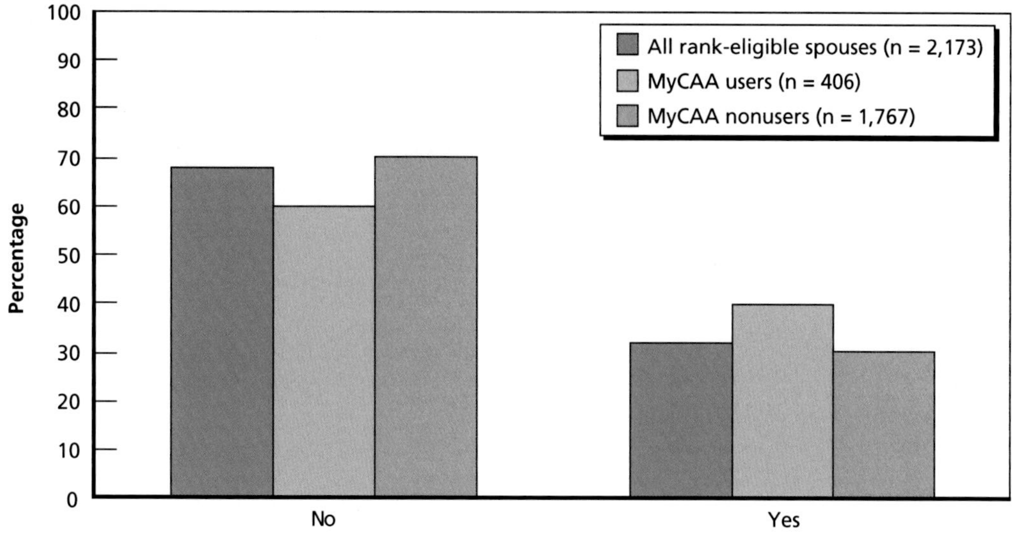

NOTE: MyCAA users differ significantly from nonusers, $p < 0.05$.
RAND *RR784-5.2*

Figure 5.3
Percentage of Spouses Who Want to Work, Among Those Who Have Not Been Looking for Work in the Last Four Weeks, by MyCAA Use

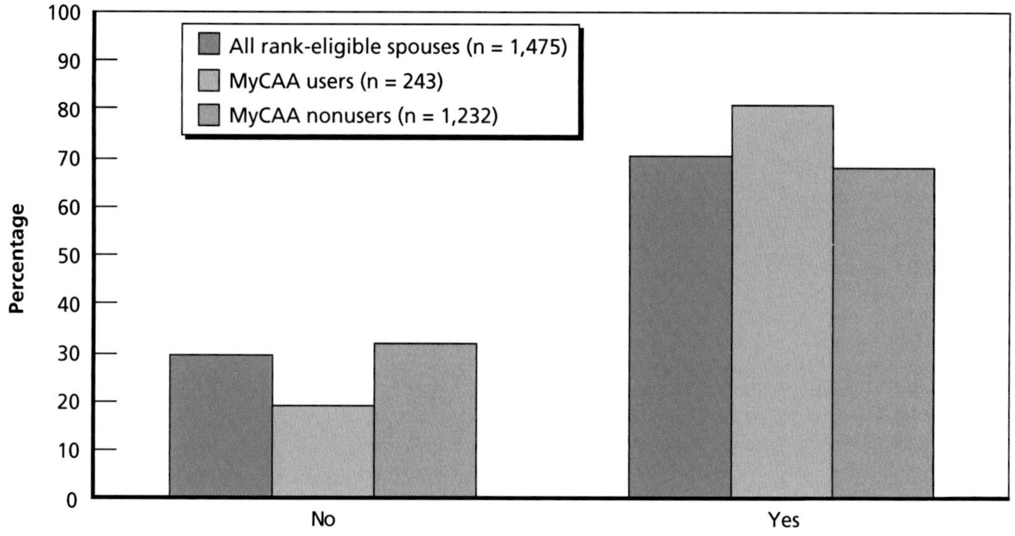

NOTE: MyCAA users differ significantly from nonusers, $p < 0.05$.
RAND *RR784-5.3*

Table 5.1
Reasons for Not Looking for Work, by MyCAA Use

	All Rank-Eligible Spouses Who Have Not Been Looking for Work During the Last Four Weeks (%)			Subset Who Want to Work (%)		
	All (n = 1,500)	MyCAA Users (n = 247)	MyCAA Nonusers (n = 1,253)	All (n = 1,010)	MyCAA Users (n = 197)	MyCAA Nonusers (n = 813)
Want to stay home to care for children	73	63[a]	75[a]	62	56	64
Child care is too costly	62	61	62	62	60	63
Do not want to work[b]	31	26	33	—	—	—
Attending school/training	27	63[a]	19[a]	29	61[a]	21[a]
[Service member] does not want me to work	26	18[a]	28[a]	11	4[a]	12[a]
Do not have child care available to me	22	22	23	25	25	25
Lack necessary work experience	21	26	20	26	27	26
Preparing for/recovering from PCS move	21	23	20	23	29	21
Lack the necessary schooling, training, or skills	21	22	21	25	23	26
Cannot find work that matches my skills	21	20	21	27	25	27
Unable to work while [service member] is deployed	21	17	22	22	20	23
Other	20	18	20	26	23	27
Not physically prepared to work (e.g., pregnant, sick, disabled)	18	21	18	19	20	19
No jobs in my career field where I live	16	14	17	21	18	21
Homeschooling children	12	10	12	10	9	10
Employed but did not work for pay or profit in past week	3	3	3	4	4	3

[a] MyCAA users and nonusers differ significantly at $p < 0.05$. The web survey participants who responded to a different set of employment questions did not receive this follow-up question and are not included in these estimates.

[b] Analysis of those who want to work (last three columns) excludes spouses who report "Do not want to work."

MyCAA Users and Nonusers Differed Significantly in Terms of Employment

MyCAA users and nonusers showed significant differences in Figures 5.1–5.3. MyCAA users were statistically significantly less likely than nonusers to be working, possibly because users were in school. MyCAA users were also significantly less likely than nonusers to be looking for work, and MyCAA users were significantly more likely than nonusers to want to be working.

What Are the Reasons for Spouses Not Looking for Work?

Spouses who were not working and not looking for work were asked a series of follow-up questions about why they were not looking for work. We examined these responses both for spouses who had not been looking for work in the last month and for the subset who wanted to work but had not sought employment. Spouses could provide multiple responses to this question.[2] Responses and corresponding percentages are reported in Table 5.1. The first set of columns displays the results for all the rank-eligible spouses in our analytic sample who were not working and had not been looking for work in the four weeks before the survey. The second set of columns in Table 5.1 depicts reasons why spouses were not looking for work for the subset of these spouses who reported wanting to work. For the most part, the reasons reported by this subset were remarkably similar to those reported by the full group of spouses who were not looking for work, but we highlight some differences in our discussion below.

Education and Child Responsibilities Were Common Reasons for Not Looking for Work

As this table indicates, spouses reported many reasons for not looking for work. One reason given for not looking for work was that they were in the midst of their education. Nonworking recent recipients of MyCAA scholarships were significantly more likely than other nonworking rank-eligible spouses to indicate that attendance at school or training was the reason for not pursuing employment (63 percent compared to 19 percent).

Child responsibilities were another commonly reported factor. A majority of spouses who were not employed reported that their main reasons for not seeking work were wanting to stay home with children (73 percent) and the expense of child care (62 percent). Even when asked to select only one main reason for not looking for work, 41 percent of this subset of spouses reported that the main reason was that they wanted to stay home with children (not shown in Table 5.1). When we look only at the subset

2 For the analysis of reasons for not working among those who want to work, we screened out spouses who do not want to work in two ways: (1) if they reported "no" to the question, "regardless of your current employment status, do you want to work" or (2) if they reported "yes" that one of the reasons they are not looking for work is because they "do not want to work."

of spouses who reported wanting to work, over 60 percent also cited wanting to stay home with children and child care costs as reasons for not looking for work. Twelve percent of spouses reported that homeschooling children kept them from looking for work.

We should also note that, compared to respondents who were recent MyCAA users, those who were not recent MyCAA users were significantly more likely to indicate that their service member did not want them to work. Given that MyCAA scholarships are provided for career-focused licenses, certificates, and associate's degrees, it makes sense that fewer MyCAA users have marital partners who object to their working.

Gap Between Spouse Qualifications and Available Jobs Also Explains Why Spouses Did Not Look for Work

Among spouses who were not working, some of the reasons for not looking for work included spouses' lack of necessary work experience (21 percent); spouses' lack of necessary schooling, training, or skills (21 percent); spouses' inability to find work that matches their skills (21 percent); and spouses' sense that there are no jobs in their career field where they currently live (16 percent).

Spouses who reported wanting to work but not looking for work were more likely to report that there were "no jobs in my field where I live" (21 percent) than the full sample of spouses not looking for work (16 percent). This challenge is one that MyCAA is specifically designed to help address: MyCAA focuses on assisting spouses with education and training that will translate into greater employability in high-growth, high-demand career fields that they can take with them wherever they live. The Military Spouse Career Center and the Military Spouse Employment Partnerships are other SECO resources established to help spouses with job searches, and the career center can help spouses with educational needs beyond the scope of MyCAA to locate financial assistance and understand how to apply for it.

Deployment and PCS Moves Were Also Reasons for Not Looking for Work

Finally, 21 percent of these spouses who were not employed reported that they had not been looking for work in the four weeks before the survey because they were either preparing for or recovering from a PCS move. The same percentage of spouses reported that they were unable to work while their service members were deployed. As we noted above, service member absences can cause a great deal of the household and family responsibilities to fall on the military spouse, limiting opportunities for employment.

What Kinds of Jobs Are Military Spouses Taking?

In addition to understanding who works and why some spouses have dropped out of the workforce, we investigated the characteristics of the jobs taken by military spouses. We looked at several characteristics of jobs using these data, including

- reports of the main reason for working part-time instead of full-time
- whether spouses are currently employed within their area of education or training
- how satisfied spouses are with their jobs (scale of 16 aspects of the workplace)
- how long it took to find employment after the last PCS move
- whether the spouse needed a new license or credential to work at the new duty location and, if so, how long it took to acquire the new license or credential.

Some Spouses Were Working Part-Time Only Because They Could Not Find Full-Time Work

Rank-eligible spouses who indicated that they were working for pay or profit but working fewer than 35 hours per week (650 of the 1,763 working spouses, or 37 percent) were asked the main reason they were working part-time instead of full-time.[3] Table 5.2 lists the main reason reported by spouses who answered that question.

Table 5.2
Main Reason Given for Working Part-Time Rather Than Full-Time, Among Spouses Working Fewer than 35 Hours a Week, by MyCAA Use

	All Rank-Eligible Spouses (n = 617) (%)	MyCAA Users (n = 116) (%)	MyCAA Nonusers (n = 501) (%)
Could find only part-time work	29	31	28
Other	19	17	20
Child care problems	11	12	11
Want to spend time with children	11	9	11
Do not want to work full-time	8	9	8
Other family/personal obligations	5	7	5
Slack work/business conditions	5	6	5
Am self-employed	6	5	6
Seasonal work	3	3	3
Health/medical limitations	1	1	1
Do not have required license or credential	2	0	2

[3] The web survey participants who responded to a different set of employment questions are not included in these estimates.

Almost one-third (29 percent) of the spouses who provided a response reported that the main reason they were working part-time was that they could not find full-time work. This is quite striking, as it indicates that, for a large percentage of spouses, working part-time is not a choice but rather a consequence of difficulty in finding other employment. In fact, only 8 percent of spouses report that the main reason they work part-time was that they did not want to work full-time. MyCAA scholarships support education with the intent of reducing unemployment and underemployment: That we do not see significant differences between recent MyCAA users could suggest that these spouses need help leveraging their newly acquired qualifications to obtain the full-time employment they desire.

Other Explanations for Part-Time Work

The desire to spend time with children and child care barriers are also factors that explain why spouses are not employed full-time. Reported reasons are quite similar for MyCAA users and nonusers and do not differ significantly.

A significant minority (19 percent) of spouses indicated that a reason other than those listed on the survey was their primary explanation for working part-time rather than full-time. It is possible that, if given a choice, some of these spouses would have indicated that they worked part-time only because they were currently in school. Indeed, of those who selected "other," 70 percent had indicated earlier on the survey that they were currently enrolled in school or training. We should also note that some of these reported reasons might be related to each other, and the percentages for some reasons might have been higher if spouses had been asked to check all options that apply. For example, spouses who did not want to work full-time because they wanted to spend time with their children might have checked both of those responses instead of just one.

MyCAA Users Were Less Likely to Be Employed in Their Field of Education and Required More Time to Find Work After a Move

As indicated in Table 5.3, MyCAA users were significantly less likely than rank-eligible nonusers to be employed in their field of education, largely because they were much more likely than nonusers to be in school at the time of the survey.[4] MyCAA users also tended to take more time than rank-eligible nonusers finding work after military moves (i.e., PCS) and took more time to acquire a new professional or occupational license after a military move.

The survey does not capture the timing of MyCAA use relative to PCS moves. Although these gaps could occur after MyCAA use, they could also be a primary motive for MyCAA program use, since the program is structured to support PCS-

[4] The web survey participants who responded to a different set of employment questions are not included in these estimates.

Table 5.3
Whether Employed Within Career Field, Job Satisfaction, and PCS Move Disruption, by MyCAA Use

		All Rank-Eligible Spouses (%)	MyCAA Users (%)	MyCAA Nonusers (%)
Currently employed within the area of education or training[a] (asked only of employed spouses, n = 1,602)	No	51	59	50
	Yes	49	41	50
Workplace satisfaction score in tertiles[b] (asked only of employed spouses, n = 1,213)	Least satisfied	38	44	36
	Middle	30	28	31
	Most satisfied	32	29	33
Time it took to find employment after last PCS move (asked only of spouses who indicated that they had ever experienced a PCS move, n = 1,527)[a]	< 1 month	13	11	14
	1 month to < 4 months	28	24	29
	4 months to < 7 months	22	24	22
	7 months to < 10 months	9	5	10
	≥ 10 months	28	35	26
Needed a new license or credential to work at the new duty location (asked only of spouses who indicated that they had ever experienced a PCS move, n = 1,575)	No	28	29	28
	Yes	72	71	72
Time to acquire a new license or credential (asked only of spouses who indicated that they had ever experienced a PCS move and that they needed a new license or credential at the new duty location, n = 439)[a]	< 1 month	14	7	16
	1 month to < 4 months	37	26	40
	4 months to < 7 months	21	30	19
	7 months to < 10 months	7	8	6
	≥ 10 months	20	29	20

NOTE: Sample sizes vary because of missing data/applicability of survey questions.

[a] MyCAA users and nonusers differ significantly at $p < 0.05$.

[b] Excludes spouses who report "not applicable" to all workplace items. Total scores range from 16–80 with a mean of 62. Least satisfied denotes scores of < 60, medium satisfaction is 60–69, and most satisfied is 70+.

related employment activities: recertification and licensing following a move, improving the ability to compete in the job market in their field, or seeking to transition to career fields with greater employment opportunities.

Chapter Summary

Although we found that the vast majority of spouses in our sample wanted to work (85 percent), there were many reasons why military spouses do not work.

Prior research has shown that concern about child care is one factor related to the employment of spouses of military service members (Lim, Golinelli, and Cho, 2007; Schwartz, Wood and Griffith, 1991; Zellman et al., 2009). We also found evidence that spouses may opt out of the workforce to care for children. In fact, a majority of spouses in the survey who were not working reported that wanting to stay at home with children and the expense of child care were the main reasons they were not seeking work. Prohibitive child care costs were also a factor reported by spouses as reasons for not seeking work, despite DoD providing income-based fees for on-installation child care and subsidies for off-installation care. Similar factors were also commonly cited as reasons for working part-time instead of full-time. As mentioned in the previous chapter, this finding could be related to lack of awareness about DoD child care options or to lack of availability.

Reported reasons for not working and working part- instead of full-time were remarkably similar for MyCAA users and nonusers. The one exception was the response of being in school, which was more commonly cited by MyCAA users as a reason for not working. We also found some differences between MyCAA users and nonusers with respect to job characteristics. For instance, MyCAA users were less likely than nonusers to be employed in their field of education but were much more likely than nonusers to be in school at the time of the survey. MyCAA users also tended to take more time than nonusers finding work after military moves and in acquiring a new professional or occupational license after a military move. It is important to note that in this cross-sectional study, it is difficult to know if these differences are consequences of being back in school and utilizing the MyCAA program or are reasons these spouses chose to pursue the MyCAA program in the first place.

CHAPTER SIX

Conclusions and Recommendations

Although RAND's analyses of a subset of military spouse responses on the 2012 ADSS do not constitute an evaluation of MyCAA program effectiveness, they do tell us that MyCAA and other DoD SECO programs are targeting some of the key education and employment hurdles facing civilian military spouses of active-duty service members in MyCAA-eligible pay grades. In this chapter, we present recommendations designed to address the issues raised in previous chapters.

Use of MyCAA

Of eligible spouses who did not use MyCAA in the year before taking the survey, more than half indicated on the survey that they were unaware of the program. Awareness of MyCAA is important to enable the program to identify and help military spouses obtain financial assistance for earning an associate's degree, license, or certificate for a portable career. Another important barrier to MyCAA use involves understanding program eligibility requirements. For MyCAA nonusers who are aware of the program, the most commonly reported barrier to use was a lack of time. To address these issues, RAND recommends the following:

- **Help spouses manage competing responsibilities so that they can benefit from MyCAA scholarships.** Career counselors at the Military Spouse Career Center might help spouses find ways to manage competing responsibilities (e.g., by linking them to child care options or connecting them to other staff at Military OneSource for assistance in locating a home repair referral source).
- **Promote MyCAA on an ongoing basis.** MyCAA's primary targets are the spouses of service members who are entry-level or very early in their careers. We found that more than half of rank-eligible nonusers were unaware of MyCAA. These unaware nonusers were much more likely to have been married less than three years than nonusers who were aware of the scholarships. Because there is a continuous flow of new spouses into the military community at the MyCAA eligibility level, promotion of MyCAA must be ongoing.

- **Coordinate with the four services to ensure that promotional activities for MyCAA target all eligible pay grades.** DoD needs to coordinate with the services to ensure that MyCAA is not mistakenly advertised as being solely for junior enlisted spouses or for spouses of "junior personnel" and to ensure that promotional activities target spouses of all eligible pay grades. To promote understanding of eligibility among spouses, DoD could also describe eligibility in terms of service-specific rank (e.g., corporal) and not just pay grade.

Higher Education

Spouses who reported wanting to be in school but who were not enrolled most commonly cited cost as a reason for not pursuing education—82 percent of rank-eligible respondents gave this reason. This demonstrates that the MyCAA program is targeting a current need by providing a way for rank-eligible spouses to address some of the costs of education. However, MyCAA scholarships do not fund the pursuit of bachelor's or graduate degrees, which can also be cost-prohibitive to spouses. Some degrees (e.g., law, accounting, veterinary medicine, childhood education) can help spouses prepare for portable careers with a high earning potential. Family responsibilities and the costs of child care were also educational hurdles shared by both rank-eligible MyCAA users and nonusers. Each was cited as a barrier by more than half of spouses who would like to be in school but were not. These issues can be addressed by the following:

- **Help spouses interested in postsecondary education identify financial aid opportunities.** Spouses may be eligible for federal financial aid from the Department of Education. Service members with more than six years of service are eligible to transfer their Post-9/11 GI Bill educational benefits to their spouses. States, educational institutions, and private foundations are other sources of financial assistance for postsecondary education. Career counselors at DoD's Military Spouse Career Center are equipped to help spouses identify financial aid opportunities beyond MyCAA for spouses of any pay grade who are interested in pursuing any type of postsecondary education.
- **Have Military Spouse Career Center career counselors inquire about child care needs.** This practice could contribute to DoD's efforts to ensure that spouses are aware of DoD-subsidized child care options and how to apply. The process for obtaining a MyCAA scholarship requires that new applicants speak with a career counselor from the Military Spouse Career Center, so DoD has the opportunity to ensure that every MyCAA recipient has been offered information about assistance with the cost of child care.
- **Investigate whether military child care options can be aligned to match the needs of spouses in school.** DoD child care availability and access is designed

primarily to meet the needs of employed spouses. DoD should consider investigating whether it would be feasible to align military child care options to also match the needs of spouses in school in terms of available hours and schedules that may shift by semester.

- **Investigate whether MyCAA partner institutions offering on-site child care would be willing to offer child care discounts, scholarships, or space priority to MyCAA recipients during their enrollment terms.** Many schools do not offer on-site child care services for students—particularly private and for-profit institutions (Miller, Gault, and Thorman, 2011)—and those that do may have waiting lists. Still, some spouses who attend schools that do offer child care might be able to benefit if discounts or scholarships could be made available or if military spouses could receive some level of priority on any waiting lists. Schools might be motivated to offer MyCAA child care discounts if MyCAA gave the schools that offer this benefit additional visibility on the MyCAA website. Perhaps some schools would be willing to offer such a benefit to any enrolled military spouse. Some colleges already offer scholarships to help student parents with the cost of child care (Miller, Gault, and Thorman, 2011) and may be willing to add "military spouse" to their eligibility criteria.
- **Through Military Spouse Career Center career counseling, explore whether online classes can help spouses with challenges in accessing education.** Spouses reported that factors that prevent them from attending school or training include frequent moves, service member deployment, lack of transportation, inconvenient school hours or locations, and conflicts with school and work schedules. Career counselors can help spouses identify programs available completely or partially online. Of course, many occupations require hands-on learning (e.g., animal trainer, electrician, chef), but some of the required coursework may be possible to obtain online, and some occupations may be particularly well-suited to online education (e.g., those in information technology).

Employment

The vast majority of spouses surveyed who were not working and not looking for work reported that they wanted to be working. However, for many spouses, school, a gap between qualifications and available jobs, and child care costs made it difficult to find suitable work or to find time to look for work. Underemployment, including the inability to find full-time work among those who would like it, is also a concern: Only 8 percent of part-time workers indicated that their main reason for working part-time was that they did not want to work full-time, and that was true of both recent users of MyCAA and nonusers. Among rank-eligible spouses with unmet employment goals, common obstacles were related to a disconnect between spouse skills and available job opportunities. MyCAA users were less likely to be employed in their area of education and took longer to find work after a PCS move than did MyCAA nonusers. These issues can be improved by the following:

- **Reach out to recent MyCAA graduates to assist with job searches.** The Military Spouse Career Center could reach out to recent MyCAA graduates to provide job search support: Its career counselors are equipped to help any military spouse with job searches, resumes, interview preparation, and connections with MSEP.
- **Connect spouses with opportunities for internships in their fields while in school.** Even before MyCAA users finish their occupational schooling or training, there are ways to improve their employment prospects. Internships might not be the right choice for all spouses or occupations or for every school term. Where appropriate, spouses should be encouraged to consider internships while in school, potentially with MSEP companies to assist in moving them into jobs more closely aligned with their current training. This offers the potential to help in a number of ways. First, spouses gain some work experience in their field (recall that 21 percent of spouses who were not employed reported that lack of necessary work experience was a reason they were for not looking for work). Second, spouses who are employed fall into the first-priority category for available space for full-time, DoD-subsidized child care, so paid internships would help address a barrier to education (child care costs). Third, internships can lead to employment with the same company, but even if they do not (the company does not have openings or the spouse moves), internship supervisors can potentially serve as job references. Regarding internships with MSEP companies, these partners have pledged to offer transferable, portable jobs so that spouses do not have to lose seniority or experience an employment gap following a PCS move. These partners include small or regional businesses, global businesses, defense contractors, universities, federal agencies, and nongovernmental organizations. Finally, there should also be continued integration of the MyCAA, Military Spouse Career Center, and MSEP programs, which could help expand internship opportunities for MyCAA users while in school.
- **Have Military Spouse Career Center career counselors inquire about callers' child care needs.** We echo our recommendation above: Leverage the moment when spouses consult with career counselors about suitable occupations, job searches, and other employment issues to offer information about DoD-subsidized child care.

Future Surveys

The 2012 ADSS contained a great deal of information useful for exploring military spouse use of the MyCAA program and their preferences, experiences, and barriers concerning higher education and employment. Should the next ADSS or another survey of military spouses focus on these topics, some additional questions on MyCAA and SECO program use would permit an even fuller exploration of their value.

First, we recommend asking MyCAA beneficiaries why they used the program so that analysts could explore whether underemployment or barriers to employment vary by MyCAA purpose. Did the spouses intend for the scholarships to help them

- prepare for their first career
- change careers
- recertify in the same career after a PCS move
- improve their pay, promotion opportunities, or job prospects within the same career field?

Our analyses could also have benefitted from more information about the timing and level of MyCAA use. Because the 2012 ADSS asked only about MyCAA use in the previous year, we were unable to distinguish spouses who ever received MyCAA benefits from those who never did. We recommend inquiring about all MyCAA use, not just use in the prior year, including how long ago spouses used it. It would be useful to know whether spouses completed their MyCAA-supported degree, certificate, or license or when they estimate they will be able to complete it. With this information, researchers could explore whether and how survey responses differ according to whether (a) spouses never used the program, (b) spouses were using the program at the time of the survey, (c) spouses had completed the program within the past year, or (d) spouses had completed the program more than one year ago. It might also be useful to link participant responses across multiple years of the ADSS so that analyses could compare spouse responses before, during, and after use of MyCAA financial assistance.

Although DoD must be mindful of survey length so it does not deter survey participation or contribute to survey break-off, the ADSS or a survey like it could be used to understand more about other SECO initiatives as well, particularly because it allows comparisons with nonusers. Questions could address awareness and use of the Military Spouses Career Center career counseling services, including whether the career counselors helped them identify portable careers that would fit their strengths and interests, find and understand the financial aid application processes, or improve their job search and application skills. Questions about MSEP could also address awareness of the program and use of the website for job seekers and employment by an MSEP partner. The survey could ask about awareness and use of the SECO website. Finally, the survey could fill a major knowledge gap regarding spouse awareness and use of unemployment compensation for trailing spouses following a military PCS, which the DoD State Liaison Office has been working on with state governments.

Directions for Future Study

This study extends the current literature, but we have also identified directions for future research. The aim of this analysis of 2012 ADSS survey data was to provide a broad overview of preferences, goals, and barriers to MyCAA use, education, and employment among spouses eligible for MyCAA based on the rank of their service member. To do so, most of our analyses are very descriptive (primarily crosstabs). Future work could provide a more in-depth analysis of many of these findings. For instance, researchers could look more closely at some of the barriers and outcomes for different demographic groups and services. Areas of the survey that could be more thoroughly explored include items on marital problems; health and well-being of the spouses, children, and service members; and deployments and postdeployment reintegration. Furthermore, it may be worth exploring whether there are enough spouses who respond to more than one ADSS to permit an analysis of how or whether individual spouses' responses vary over time (a longitudinal analysis). Another approach would be to identify who is still a military spouse five years after the survey was conducted, whether any spouse survey responses could help predict which service members reenlist or remain in the military after their first term of service, or which spouses remain married to their service member. This study took an important step in identifying the preferences, needs, and characteristics of MyCAA users and nonusers who may be eligible for MyCAA, but more work is needed to better understand who uses these programs, why, and to what end, and what changes can better facilitate use for spouses who wish to improve their education and career opportunities.

This study focused on military spouse education explicitly and not just as a control variable for income or employment. Consistent with past work on military spouses (Harrell, 2000) and American adults in general (Perna and Jones, 2013), we found that educational and child care costs prevent spouses from pursuing additional schooling, and that many spouses who wish to be in school are not. Although the 2012 ADSS advances our knowledge about military spouses' educational preferences and barriers to reaching their goals, there is a lot we do not know about their educational choices and the factors that could slow or prevent the completion of degrees and certificates. For example:

- What criteria do spouses consider when selecting a major, an educational goal, and an institution of higher learning?
- What are the average times to completion for spouses pursuing different educational goals and how do those compare to national averages?

- Are spouses taking relatively efficient paths toward their degrees and certificates, or, like many American students, are they spending time and money taking a lot of unnecessary courses (National Governor's Association Common Completion Metrics, 2011)? Or how many spouses are taking college classes for other reasons, such as the love of learning, fitness and recreation, or gaining specific skills (e.g., photography, foreign language, creative writing)?
- What is the level of need for remedial classes; tutoring; writing, test-taking, and study skills support; and English as a Second Language courses? How many spouses are coping with learning disabilities?
- Do spouses' educational trajectories tend to get derailed during critical transition points, such as after PCS moves or when transferring from a two-year to a four-year institution?

More scholarly attention to higher education among military spouses is needed to be able to answer these questions. The results may not only enrich our understanding of military spouse education but may also point to new ways that DoD could help support spouses.

Why do some military spouses not use programs and services for which they are eligible and have a need? The results of this study echo prior reports of concerns that lack of program awareness may keep some individuals from utilizing military programs (Harrell, 2000; Miller et al., 2011a; GAO, 2012b, pp. 10–11). This study also found that spouses who have been married less than three years or are married to junior enlisted personnel are less likely than other rank-eligible spouses to be aware of MyCAA. Spouses who are aware of MyCAA are less likely to be living in civilian housing than spouses who are unaware.

There is no standardized process for orienting new military spouses to the military organization, culture, programs, and services. Additional research could explore new spouses' early experiences as military spouses, the methods that are most successful at reaching different groups of spouses (e.g., those working full-time, those living off base, those with little to no prior exposure to the military), and ways to ensure that they have the information they need or that they know where to find it. Research could also explore whether or how these early experiences relate to military family problems and unmet needs, spouse employment and earnings, spouse satisfaction with military life, and service member retention.

Concluding Remarks

In sum, among spouses with unrealized educational goals, the costs of education, family responsibilities, and the costs of child care were the most commonly reported barriers.

Among spouses with unmet employment goals, common obstacles were related to a disconnect between spouse skills and available job opportunities.

Military spouse feedback through the 2012 ADSS, such as that provided here, can help DoD adjust its offerings to better assist the spouses' education and employment trajectories. For example, DoD could take additional steps to ensure that MyCAA program eligibility is clear to potential users and might evaluate whether military child care options sufficiently meet the needs of spouses in school. DoD might also consider expanding internship opportunities for MyCAA users and provide job search support for MyCAA graduates. Developing and expanding such programs could be instrumental in helping military spouses advance and achieve their career goals.

APPENDIX A

Demographic, Family, and Military Characteristics of Spouses Included in This Study

To capture service members' characteristics, we included information on a number of factors, including service members' unit and pay grade, and spouses' age, level of education, family status, reported financial condition, and physical and mental health.

Service Members' Characteristics

First, we looked at service units, which were matched from administrative records. These include Army, Navy, Marine Corps, and Air Force. We also examined pay grade categories that were coded as E-1 to E-5, W-1 to W-2, and O-1 to O-2. Finally, to capture information on the amount of time the service member is away from home, we included information on the number of nights away from home in previous three years, which were coded as 0, 1–59, 60–179, 180–269, 270–449, 450–1,095, and missing, if information was not reported.

Spouses' Characteristics

Spouses' characteristics include age coded in five categories: < 26 years old, 26–30, 31–35, 36–40, and > 40; gender; a dichotomous measure of race-ethnicity capturing whether the spouse is white or nonwhite; and education, categorized as less than 12 years of school (no diploma), high school graduate (diploma or equivalent), some college (no degree), associate's degree (e.g., AA, AS), bachelor's degree (e.g., BA, AB, BS), and master's degree or more.

Family Characteristics

Family characteristics include number of years married, categorized as < 3, 3–6, 6–10, 10–15, and 15+; whether there are any children in the family and children's age groups, coded as no children living at home, children < 6, children 6–13 , and children 14–18;

whether the family is enrolled in the EFMP; and housing, namely, whether the family lives on base, off base in military housing, or off base in civilian housing.

Self-Reported Financial Circumstances, Health, and Satisfaction with the Military

In addition to characteristics of service members, spouses, and families, we also examined self-reported financial circumstances, health, and satisfaction with the military.

Financial Circumstances

To capture financial circumstances, we looked at several factors. The first was spouse's monthly income in categories, including no income, more than 0 income but < $1,000, $1,000 – < $2,000, $2,000 – < $3,000, $3,000 – < $5,000, $5,000 – < $7,000, ≥ $7,000, and missing income information. We also examined a measure of savings, namely, whether the respondent reported $500 or more in emergency savings. For this variable, responses of "yes" were coded 1 and "no" or "don't know" were coded 0. Reported financial condition was captured from a question asking, "Which best describes the financial condition of you and your spouse? Mark one." Although five response options were provided, we used the DMDC recoded variable with three categories, specifically whether a spouse reports being (1) very comfortable and secure/able to make ends meet without much difficulty, (2) occasionally have some difficulty making ends meet, (3) tough to make ends meet but keeping our heads above water/in over our heads.

The 2012 ADSS data also included an indicator variable for whether the respondent reported that at least one financial problem happened to them or their spouse in the last 12 months. The list of possible problems included

- bounced two or more checks
- failed to make a monthly/minimum payment on credit card, AAFES, NEXCOM account, or Military Star Card account
- fell behind in paying rent or mortgage
- was pressured to pay bills by stores, creditors, or bill collectors
- had telephone, cable, or Internet shut off
- had water, heat, or electricity shut off
- had a car, household appliance, or furniture repossessed
- failed to make a car payment
- filed for personal bankruptcy
- had to pay overdraft fees to a bank or credit union two or more times.

Health

The survey also solicited self-reports of physical and mental health. Physical health items on the 2012 ADSS came from the general health perceptions subscale on the Short-Form Health Survey (SF-36) of the Medical Outcomes Study questionnaire (Ware and Sherbourne, 1992). It asked: "How true or false is each of the following statements for you? Mark one answer for each statement." Response categories ranged from "definitely true" to "definitely false." Items included

- I am as healthy as anybody I know.
- I seem to get sick a little easier than other people.
- I expect my health to get worse.
- My health is excellent.

We calculated a total score as the average across the four items. Scores ranged from 1–4, with items reverse-coded so that 4 indicates the most healthy.

Mental health items on the 2012 ADSS came from the PHQ-4, which has been validated in several prior studies (Kroenke, Spitzer, and Williams, 2003; Kroenke et al., 2009; Löwe et al., 2010; Mitchell and Coyne, 2007). The dataset RAND received from DMDC included a mental health score calculated using spouses' responses to several items provided in response to the question "Over the last two weeks, how often have you been bothered by any of the following problems?" Response categories ranged from "nearly every day" to "not at all." Problems included little interest or pleasure in doing things; feeling down, depressed, or hopeless; feeling nervous, anxious, or on edge; and not being able to stop or control worrying. Scores again ranged from 1–4, with 4 representing those who are the most healthy.

In our analyses, we examine physical and mental health scores as both linear measures and as dichotomous indicators of mental and physical health capturing whether health is better than that of the mean reported by the sample of spouses. Results are consistent regardless of how the measure is constructed. The dichotomized version (better or worse than the mean) is presented in the tables in the text.

Our final measure of health captures personal stress and was coded in response to a question asking, "Overall, how would you rate the current level of stress in your personal life?" Responses included "much less than usual," "less than usual," "about the same as usual," "more than usual," and "much more than usual." We recoded this variable to capture (1) more and much more than usual stress and (2) about the same or less than usual.

Military Satisfaction

We also look at two measures capturing reported military satisfaction. The first was captured in response to the question "Overall, how satisfied are you with the military way of life?" Responses ranged from "very satisfied" to "very dissatisfied." Spouses reporting that they are satisfied or very satisfied were coded 1; other responses were coded 0. The second measure was coded in response to the question "Do you think your spouse should stay on or leave active duty?" Spouses who report that they favor or strongly favor staying were coded as favoring staying with the military.

APPENDIX B

Comparison of MyCAA Nonusers Who Were Aware of the Scholarships to Unaware Nonusers

Table B.1
Demographic Characteristics of Spouses Who Did Not Use MyCAA in the Previous Year, by Awareness of MyCAA

		Unaware of MyCAA (%)	Aware of MyCAA (%)
Age (years)[a]	< 26	38	36
	26–30	34	37
	31–35	15	18
	36–40	8	5
	> 40	5	4
Gender[a]	Female	90	96
	Male	10	4
Race-ethnicity	White	64	67
	Nonwhite	36	33
Education[a]	< 12 years of school (no diploma)	3	2
	High school graduate (diploma or equivalent)	16	13
	Some college (no degree)	38	36
	Associate's degree (e.g., AA, AS)	15	16
	Bachelor's degree (e.g., BA, AB, BS)	22	26
	Master's degree or more	6	7
Vocational certificate	Received vocational or technical diploma	19	17
	Did not receive	81	83
Own monthly income	None	31	34
	> 0 but ≤ $1,000	13	15
	> $1,000–≤ $2,000	14	13
	> $2,000–≤ $3,000	12	12
	> $3,000–≤ $5,000	13	12
	> $5,000–≤ $7,000	3	3
	> $7,000	2	1
	Missing income information	12	11

NOTE: Sample sizes vary slightly because of missing data.

[a] MyCAA users differ significantly from MyCAA nonusers at $p < 0.05$ based on chi^2 test.

Table B.2
Spouse's Service Member Characteristics for Spouses Who Did Not Use MyCAA in the Previous Year, by Spouse Awareness of MyCAA

		Unaware of MyCAA (%)	Aware of MyCAA (%)
Service[a]	Army	42	47
	Navy	22	17
	Marine Corps	15	15
	Air Force	22	21
Pay grade[a]	Junior enlisted (E-1 to E-4)	58	51
	Noncommissioned officer (E-5)	34	40
	Warrant officer (W-1 to W-2)	2	3
	Junior officer (O-1 to O-2)	6	6
Nights in the last 36 months service member was away from home	0	5	5
	1–59	18	16
	60–179	17	17
	180–269	14	14
	270–449	24	26
	450–1,095	19	19
	Missing	3	3

NOTE: Sample sizes vary slightly because of missing data.

[a] MyCAA users differ significantly from MyCAA nonusers at $p < 0.05$ based on chi^2 test.

Table B.3
Family Characteristics of Spouses Who Did Not Use MyCAA in the Previous Year, by Awareness of MyCAA

		Unaware of MyCAA (%)	Aware of MyCAA (%)
Years married[a]	< 3	43	28
	3–6	31	39
	6–10	17	22
	10–15	6	9
	15+	3	2
Children[a]	None living at home	40	35
	Age < 6	37	36
	Age 6–13	18	23
	Age 14–18	6	5
Enrolled in the EFMP	Yes	9	11
	No (or not applicable)	91	89
Housing[a]	On base	28	32
	Military off base	9	10
	Civilian housing	64	58

NOTE: Sample sizes vary slightly due to missing data.

[a] MyCAA users differ significantly from MyCAA nonusers at $p < 0.05$ based on chi^2 test.

APPENDIX C

Results of Logistic Regression Models Comparing MyCAA Users to Nonusers on Demographic, Family, and Military Characteristics

Table C.1
Odds Ratios from Logistic Regression Models of the Association Between Demographic, Family, and Military Characteristics and Recent MyCAA Use

		Model 1: All Rank-Eligible Spouses (n = 3,913)	Model 2: Users and Aware Nonusers Only (n = 2,163)
Age (years)	< 26	1.29	0.93
	26–30	1.19	0.97
	31–35	0.64	0.54
	36–40	0.85	0.97
	> 40	reference	reference
Gender	Female	2.58***	1.43
Race-ethnicity	White	0.71***	0.67***
Education	< 12 years of school (no diploma)	0.10**	0.10**
	High school graduate (diploma or equivalent)	0.41***	0.42***
	Some college, no degree	reference	reference
	Associate's degree (e.g., AA, AS)	0.69**	0.68**
	Bachelor's degree (e.g., BA, AB, BS)	0.42***	0.38***
	Master's degree or more	0.42***	0.39***
Vocational/technical diploma		1.58***	1.58***
Years married	< 3	reference	reference
	3–6	1.46**	1
	6–10	1.02	0.68*
	10–15	1.12	0.7
	>15	0.84	0.72

Table C.1—Continued

		Model 1: All Rank-Eligible Spouses (n = 3,913)	Model 2: Users and Aware Nonusers Only (n = 2,163)
Children	None living at home	reference	reference
	Age < 6	0.69**	0.75*
	Age 6–13	1.02	0.94
	Age 14–18	1.21	1.15
Service	Army	reference	reference
	Navy	0.73*	0.91
	Marine Corps	0.9	0.99
	Air Force	0.74*	0.86
Pay grade	E-1 to E-4	reference	reference
	E-5	0.9	0.88
	W-1 to W-2	1.04	1.11
	O-1 to O-2	0.71	0.72
Enrolled in the EFMP	Yes	0.9	0.88
Living arrangements	On base	reference	reference
	Military off base	0.79	0.77
	Civilian housing	0.80*	0.86
Nights in the last 36 months service member away from home	0	reference	reference
	1–59	0.78	0.87
	60–179	0.9	0.96
	180–269	0.87	0.94
	270–449	0.81	0.88
	450–1,095	0.8	0.85
	Missing	0.84	0.71
More stress than usual in personal life		1.11	1.02
Below average mental health		1	0.94
Below average physical health		0.9	0.85
Own monthly income	None	1.30**	1.27*
	Some	reference	reference
	Missing income information	1.13	1.02

Table C.1—Continued

		Model 1: All Rank-Eligible Spouses (n = 3,913)	Model 2: Users and Aware Nonusers Only (n = 2,163)
Financial condition	Very comfortable	reference	reference
	Occasionally have some difficulty	0.99	1.01
	Difficulty making ends meet	1	1.19
At least one bad outcome		0.99	1.02
Savings > $500		0.88	0.88
Satisfied with the military		0.95	0.86
Favor staying in the military		1.01	0.96
Completed mail survey		0.97	0.84

* $p < 0.05$, ** $p < 0.01$, *** $p < 0.001$.

References

Barnett, Rosalind Chait, Karen C. Gareis, and Robert T. Brennan, "Wives' Shift Work Schedules and Husbands' and Wives' Well-Being in Dual-Earner Couples with Children: A Within-Couple Analysis," *Journal of Family Issues*, Vol. 29, No. 3, 2008, pp. 396–422.

Booth, Bradford, "Contextual Effects of Military Presence on Women's Earnings," *Armed Forces & Society*, Vol. 30, No. 1, 2003, pp. 25–51.

Booth, Bradford, William W. Falk, David R. Segal, and Mady Wechsler Segal, "The Impact of Military Presence in Local Labor Markets on the Employment of Women," *Gender & Society*, Vol. 14, No. 2, 2000, pp. 318–332

Booth, Bradford, Mady Wechsler Segal, and D. Bruce Bell, *What We Know About Army Families: 2007 Update*, Fairfax, Va.: ICF International, 2007.

Castaneda, Laura Werber, and Margaret C. Harrell, "Military Spouse Employment: A Grounded Theory Approach to Experiences and Perceptions," *Armed Forces & Society*, Vol. 34, No. 3, 2008, pp. 389–412.

Cooke, Thomas J., and Karen Speirs, "Migration and Employment Among the Civilian Spouses of Military Personnel," *Social Science Quarterly*, Vol. 86, No. 2, 2005, pp. 343–355.

Cooney, Richard, *Moving with the Military: Race, Class and Gender Differences in the Employment Consequences of Tied Migration*, dissertation, University of Maryland at College Park, 2003.

Cooney, Richard, Karin De Angelis, and Mady W. Segal, "Moving with the Military: Race, Class, and Gender Differences in the Employment Consequences of Tied Migration," *Race, Gender and Class*, Vol. 18, No. 1–2, 2011, pp. 360–384.

Defense Manpower Data Center (DMDC), *Survey of Active Duty Spouses: Statistical Methodology Report*, Alexandria, Va., 2013a.

———, *2012 Survey of Active Duty Spouses: Tabulations of Responses*, Alexandria, Va., 2013b.

Groves, Robert M., "Nonresponse Rates and Nonresponse Bias in Household Surveys," *Public Opinion Quarterly*, Vol. 70, No. 4, 2006, pp. 646–675.

Harkin, Tom, *Benefitting Whom? For-Profit Education Companies and the Growth of Military Educational Benefits*, Washington, D.C.: U.S. Senate Health Education, Labor and Pensions Committee, December 8, 2010.

Harrell, Margaret C., *Invisible Women: Junior Enlisted Army Wives*, Santa Monica, Calif.: RAND Corporation, MR-1223, 2000. As of November 6, 2014:
http://www.rand.org/pubs/monograph_reports/MR1223.html

Harrell, Margaret C., Nelson Lim, Laura Werber Castaneda, and Daniela Golinelli, *Working Around the Military: Challenges to Military Spouse Employment and Education,* Santa Monica, Calif.: RAND Corporation, MG-196-OSD, 2004. As of November 6, 2014:
http://www.rand.org/pubs/monographs/MG196.html

Heaton, Paul, and Heather Krull, *Unemployment Among Post-9/11 Veterans and Military Spouses After the Economic Downturn,* Santa Monica, Calif.: RAND Corporation, OP-376, 2012. As of November 6, 2014:
http://www.rand.org/pubs/occasional_papers/OP376.html

Hisnanick, John J., and Roger D. Little, "'Honey, I Love You, but . . .' Investigating the Causes of the Earnings Penalty of Being a Tied-Migrant Military Spouse," *Armed Forces & Society,* January 10, 2014. As of July 23, 2014:
http://afs.sagepub.com/content/early/2014/01/01/0095327X13512620

Hosek, James R., Beth J. Asch, C. Christine Fair, Craig W. Martin, and Michael G. Mattock, *Married to the Military: The Employment and Earnings of Military Wives Compared with Those of Civilian Wives*, Santa Monica, Calif.: RAND Corporation, MR-1565-OSD, 2002. As of November 6, 2014:
http://www.rand.org/pubs/monograph_reports/MR1565.html

Johnson, Timothy P., and Joseph S. Wislar, "Response Rates and Nonresponse Error in Surveys," *Journal of the American Medical Association,* Vol. 307, No. 17, 2012, pp. 1805–1806.

Kniskern, Mary K., and David R. Segal, *Mean Wage Differences Between Civilian and Military Wives,* Center for Research on Military Organization, University of Maryland at College Park, November 15, 2010.

Kroenke, Kurt, Robert L. Spitzer, and Janet B. Williams, "The Patient Health Questionnaire-2: Validity of a Two-Item Depression Screener," *Medical Care,* Vol. 41, 2003, pp. 1284–1292.

Kroenke, Kurt, Robert L. Spitzer, Janet B. Williams, and Bernd Löwe, "An Ultra-Brief Screening Scale for Anxiety and Depression: The PHQ-4," *Psychosomatics,* Vol. 50, No. 6, 2009, pp. 613–621.

Lim, Nelson, and Daniela Golinelli, *Monitoring Employment Conditions of Military Spouses*, Santa Monica, Calif.: RAND Corporation, TR-324-OSD, 2006. As of November 6, 2014:
http://www.rand.org/pubs/technical_reports/TR324.html

Lim, Nelson, Daniela Golinelli, and Michelle Cho, *"Working Around the Military" Revisited: Spouse Employment in the 2000 Census Data*, Santa Monica, CA: RAND Corporation, MG-566-OSD, 2007. As of November 6, 2014:
http://www.rand.org/pubs/monographs/MG566.html

Little, Roger D., and John J. Hisnanick, "The Earnings of Tied-Migrant Military Husbands," *Armed Forces & Society,* Vol. 33, No. 4, 2007, pp. 547–570.

Löwe, Bernd, Inka Wahl, Matthias Rose, Carsten Spitzer, Heide Glaesmer, Katja Wingenfeld, Antonius Schneider, and Elmar Brahler, "A 4-Item Measure of Depression and Anxiety: Validation and Standardization of the Patient Health Questionnaire-4 (PHQ-4) in the General Population," *Journal of Affective Disorders,* Vol. 122, No. 1, 2010, pp. 86–95.

Lucal, Mary Elizabeth, *Opting Out: Non-Participation in an Educational Assistance Program*, Ph.D. dissertation, Knoxville: University of Tennessee, 2012. As of November 6, 2014:
http://trace.tennessee.edu/utk_graddiss/1538

Miller, Kevin, Barbara Gault, and Abby Thorman, *Improving Child Care Access to Promote Postsecondary Success Among Low-Income Parents,* Washington, D.C.: Institute for Women's Policy Research, 2011.

Miller, Laura L., Sarah O. Meadows, Lawrence M. Hanser, and Stephanie Taylor, *Year of the Air Force Family: 2009 Survey of Active-Duty Spouses*, Santa Monica, Calif.: RAND Corporation, TR-879-AF, 2011a. As of November 6, 2014:
http://www.rand.org/pubs/technical_reports/TR879.html

Miller, Laura L., Bernard D. Rostker, Rachel M. Burns, Dionne Barnes-Proby, Sandraluz Lara-Cinisomo, and Terry R. West, *A New Approach for Assessing the Needs of Service Members and Their Families,* Santa Monica, Calif.: RAND Corporation, MG-1124-OSD, 2011b. As of November 6, 2014:
http://www.rand.org/pubs/monographs/MG1124.html

Mitchell, Alex J., and James C. Coyne, "Do Ultra-Short Screening Instruments Accurately Detect Depression in Primary Care? A Pooled Analysis and Meta-Analysis of 22 Studies," *British Journal of General Practice,* Vol. 57, 2007, pp. 144–151.

National Governor's Association Common Completion Metrics, *Time Is the Enemy,* Washington, D.C.: Complete College America, 2011.

Needels, Karen, and Heather Hesketh Zaveri, *The Implementation Evaluation of the Military Spouse Career Advancement Accounts Demonstration (2009): Part II: Additional Findings from the Military Spouse Career Advancement Accounts Demonstration: Implementation Progress and Participants' Characteristics and Plans,* Princeton, N.J.: Mathematica Policy Research, Inc., 2009.

Perna, Laura W., and Anthony P. Jones, *The State of College Access and Completion: Improving College Success for Students from Underrepresented Groups,* New York: Routledge, 2013.

Rostker, Bernard D., *I Want You! The Evolution of the All-Volunteer Force,* Santa Monica, Calif.: RAND Corporation, MG-265-RC, 2006. As of November 6, 2014:
http://www.rand.org/pubs/monographs/MG265.html

———, *America Goes to War: Managing the Force During Times of Stress and Uncertainty,* Santa Monica, Calif.: RAND Corporation, MG-380-OSD, 2007. As of November 6, 2014:
http://www.rand.org/pubs/monographs/MG380.html

Schwartz, J. Brad, Lisa L. Wood, and Janet D. Griffith, "The Impact of Military Life on Spouse Labor Force Outcomes," *Armed Forces and Society,* Vol. 17, No. 3, 1991.

SECO—*See* Spouse Education and Career Opportunities.

Segal, Mady Wechsler, "The Military and the Family as Greedy Institutions," *Armed Forces & Society,* Vol. 13, No. 1, 1986, pp. 9–38.

Spouse Education and Career Opportunities, "Career Search," 2013. As of July 19, 2014:
https://aiportal.acc.af.mil/mycaa/Resources/CareerSearch.aspx

Strazdins, L., M. Shipley, M. Clements, L. V. O'Brien, and D. H. Broom, "Job Quality and Inequality: Parents' Jobs and Children's Emotional and Behavioural Difficulties," *Social Science & Medicine,* Vol. 70, No. 12, 2010, pp. 2052–2060.

U.S. Department of Defense, *2012 Demographics Profile of the Military Community,* Washington, D.C.: Office of the Deputy Under Secretary of Defense, 2013. As of November 21, 2014:
http://www.militaryonesource.mil/12038/MOS/Reports/2012_Demographics_Report.pdf

———, Department of Defense Instruction 6060.02, *Child Development Programs,* August 5, 2014.

U.S. Government Accountability Office (GAO), *Military Child Care: DoD Is Taking Actions to Address Awareness and Availability Barriers,* Washington, D.C., 2012a.

———, *Military Spouse Employment Programs: DoD Can Improve Guidance and Performance Monitoring,* Washington, D.C., 2012b.

Ware, J. E., and C. D. Sherbourne, "The MOS 36-Item Short Form Health Survey (SF-36): I. Conceptual Framework and Item Selection," *Medical Care,* Vol. 30, 1992, pp. 473–483.

White House, *Strengthening Our Military Families: Meeting America's Commitment*, Washington, D.C., 2011.

Zaveri, Heather Hesketh, Caterina Pisciotta, and Linda Rosenberg, *The Implementation Evaluation of the Military Spouse Career Advancement Accounts Demonstration (2009): Part I: Early Implementation of the Military Spouse Career Advancement Accounts Demonstration*, Princeton, N.J.: Mathematica Policy Research, Inc., 2009.

Zellman, Gail L., Susan M. Gates, Joy S. Moini, and Marika Booth, "Meeting Family and Military Needs Through Military Child Care," *Armed Forces & Society,* Vol. 35, No. 3, 2009, pp. 437–459.